# Modern Differential Geometry for Physicists

## World Scientific Lecture Notes in Physics

*Published*

Vol. 47: Some Elementary Gauge Theory Concepts
*H M Chan and S T Tsou*

Vol. 48: Electrodynamics of High Temperature Superconductors
*A M Portis*

Vol. 49: Field Theory, Disorder and Simulations
*G Parisi*

Vol. 50: The Quark Structure of Matter
*M Jacob*

Vol. 51: Selected Topics on the General Properties of Quantum Field Theory
*F Strocchi*

Vol. 52: Field Theory: A Path Integral Approach
*A Das*

Vol. 53: Introduction to Nonlinear Dynamics for Physicists
*H D I Abarbanel, et al.*

Vol. 54: Introduction to the Theory of Spin Glasses and Neural Networks
*V Dotsenko*

Vol. 55: Lectures in Particle Physics
*D Green*

Vol. 56: Chaos and Gauge Field Theory
*T S Biro, et al.*

Vol. 57: Foundations of Quantum Chromodynamics (2nd edn.)
*T Muta*

Vol. 59: Lattice Gauge Theories: An Introduction (2nd edn.)
*H J Rothe*

Vol. 60: Massive Neutrinos in Physics and Astrophysics
*R N Mohapatra and P B Pal*

Vol. 61: Modern Differential Geometry for Physicists (2nd edn.)
*C J Isham*

Vol. 62: ITEP Lectures on Particle Physics and Field Theory (In 2 Volumes)
*M A Shifman*

Vol. 64: Fluctuations and Localization in Mesoscopic Electron Systems
*M Janssen*

Vol. 65: Universal Fluctuations: The Phenomenology of Hadronic Matter
*R Botet and M Ploszajczak*

Vol. 66: Microcanonical Thermodynamics: Phase Transitions in "Small" Systems
*D H E Gross*

Vol. 67: Quantum Scaling in Many-Body Systems
*M A Continentino*

Vol. 69: Deparametrization and Path Integral Quantization of Cosmological Models
*C Simeone*

World Scientific Lecture Notes in Physics - Vol. 61

# MODERN DIFFERENTIAL GEOMETRY FOR PHYSICISTS

Second Edition

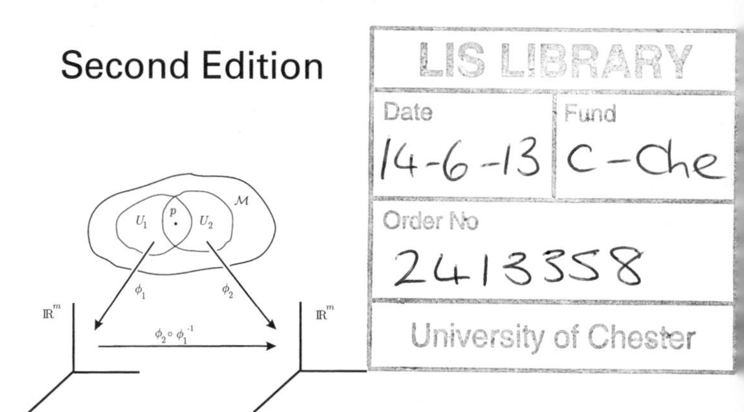


Chris J Isham

Theoretical Physics Group
Imperial College of Science, Technology and Medicine, UK

World Scientific

NEW JERSEY • LONDON • SINGAPORE • BEIJING • SHANGHAI • HONG KONG • TAIPEI • CHENNAI

*Published by*

World Scientific Publishing Co. Pte. Ltd.
5 Toh Tuck Link, Singapore 596224
*USA office:* 27 Warren Street, Suite 401-402, Hackensack, NJ 07601
*UK office:* 57 Shelton Street, Covent Garden, London WC2H 9HE

**Library of Congress Cataloging-in-Publication Data**
Isham, C. J.
Modern differential geometry for physicists / Chris J. Isham. --
2nd ed.
p. cm. -- (World Scientific lecture notes in physics : vol. 61)
Includes bibliographical references and index.
ISBN-13 978-981-02-3555-0 (alk. paper)
ISBN-10 981-02-3555-0 (alk. paper).
ISBN-13 978-981-02-3562-8 (pbk. : alk. paper)
ISBN-10 981-02-3562-3 (pbk. : alk. paper)
1. Geometry, Differential. I. Title. II. Series.
QA641.I84 1999
516.3'6--dc21 98-53245
CIP

**British Library Cataloguing-in-Publication Data**
A catalogue record for this book is available from the British Library.

First published 1999
Reprinted 2001, 2003, 2005, 2008, 2012, 2013



Printed in Singapore by World Scientific Printers.

# Preface

This book is based on lecture notes for the introductory course on modern, coordinate-free differential geometry which is taken by our first-year theoretical physics PhD students, or by students attending the one-year MSc course "Fundamental Fields and Forces" at Imperial College.

The course is concerned entirely with the mathematics itself, although the emphasis and detailed topics have been chosen with an eye on the way in which differential geometry is applied to theoretical physics these days. Such applications include not only the traditional area of general relativity, but also the theory of Yang-Mills fields, non-linear sigma models, superstring theory, and other types of non-linear field systems that feature in modern elementary particle theory and quantum gravity.

The course is in four parts dealing with, respectively, (i) an introduction to general topology; (ii) introductory coordinate-free differential geometry; (iii) geometrical aspects of the theory of Lie groups and Lie group actions on manifolds; and (iv) the basic ideas of fibre bundle theory.

The first chapter contains a short introduction to general topology with the aim of providing the necessary prerequisites for the later chapters on differential geometry and fibre bundle theory. The treatment is a little idiosyncratic in so far as I wanted to emphasise certain algebraic aspects of topology that are not normally mentioned in introductory mathematics texts but which are of potential interest and importance in the use of topology in theoretical physics.

The second and third chapters contain an introduction to differential geometry proper. In preparing this part of the text, I was particularly conscious of the difficulty which physics graduate students often experience when being exposed for the first time to the rather abstract ideas of differential geometry. In particular, I have laid considerable stress on the basic ideas of 'tangent space structure', which I develop from several different points of view: some geometric, some more algebraic. My experience in teaching this subject for a number of years is that a firm understanding of the various ways of describing tangent spaces is the key to attaining a grasp of differential geometry that goes beyond just a superficial acquiescence in the jargon of the subject. I have not included any material on Riemannian geometry as this aspect of the subject is well covered in many existing texts on differential geometry and/or general relativity.

Chapter four is concerned with the theory of Lie groups, and the action of Lie groups on differentiable manifolds. I have tried here to emphasise the geometrical foundations of the connection between Lie groups and Lie algebras, but the latter subject is not treated in any detail and readers not familiar with this topic should supplement the text at this point.

The theory of fibre bundles is introduced in chapter five, with a treatment that emphasises the theory of principle bundles and their associated bundles. The final chapter contains an introduction to the theory of connections and their use in Yang-Mills theory. This is fairly brief since many excellent introductions to the subject aimed at physicists have been published in recent years, and there is no great point in replicating that material in detail here.

The second edition of this book differs from the first mainly by the addition of the chapter on general topology; it has also been completely reset in LaTeX, thus allowing for a more extensive index. In addition, I have taken the opportunity to correct misprints in the original text, and I have included a few more worked examples. A number of short explanatory remarks have been added in places where readers and students have suggested that it might be helpful: I am most grateful to all those who drew my attention to such deficiencies in the original text. However, I have resisted the attention to add substantial amounts of new material—other than the chapter on topology—since I wanted to retain the flavour of the original as bona fide lecture *notes* that could reasonably be read in their entirety by a student who sought an overall introduction to the subject.

*Chris Isham*

Imperial College, June 1998

# Contents

# Chapter 1

# An Introduction to Topology

## 1.1 Preliminary Remarks

### 1.1.1 Remarks on differential geometry

A physics student is likely to first encounter the subject of differential geometry in a course on general relativity, where spacetime is represented mathematically by a four-dimensional differentiable manifold. However, this is far from being the only use of differential geometry in physics. For example, the Hamiltonian and Lagrangian approaches to classical mechanics are best described in this way; and the use of differential geometry in quantum field theory has increased steadily in recent decades—for example, in canonical quantum gravity, superstring theory, the non-linear $\sigma$-model, topological quantum field theory, and Yang-Mills theory.

Evidently, no excuse is needed for teaching a course on differential geometry to postgraduate students of theoretical physics. However, the impression of the subject gained from, say, an undergraduate course in general relativity can be rather misleading when viewed from the perspective of modern mathematics. Such courses usually employ a very coordinate-based approach to the subject, with little reference to the fact that more than one coordinate system may be needed to cover a manifold. In particular, although there are usually copious discussions of the effects on tensorial objects of changing from one coordinate system to another, only rarely is it emphasised that the domains of two coordinate systems may differ, and that—for example—the familiar expression involving Jacobian transformations is really only valid on the intersection of the domains of the coordinate systems concerned.

What is neglected in such approaches to differential geometry is the fact that the *topology* of a manifold may be different from that of a vector space, and hence—in particular—it cannot be covered by a single coordinate system. The modern approach to differential geometry is very different: although coordinate systems have an important role to play, the key concepts are developed in a way that is manifestly independent of any specific reference to coordinates. Concomitantly, the fact that a manifold is actually a special type of topological space becomes of greater importance, and for this reason it is appropriate to begin any text dealing with modern 'coordinate-free' differential geometry with an introduction to general topology and associated ideas. In fact, the subject of topology proper is of considerable significance in many areas of modern theoretical physics, and is well worth studying in its own right.

### 1.1.2 Remarks on topology

The subject of topology can be approached in a variety of ways. At the most abstract level, a 'topology' on a set $X$ consists of a collection of subsets of $X$—known as the *open sets* of the topology—that satisfy certain axioms (they are listed in Theorem 1.3). This special collection of subsets is then used to give a purely set-theoretic notion of characteristic topological ideas such as 'nearness', 'convergence of a sequence', 'continuity of a function' *etc.* From a physical perspective, one could say that topology is concerned with the relation between points and 'regions': in particular, open sets are what 'real things' can exist in.

Many excellent books on topology take an abstract approach from the outset[1]. However, on a first encounter with the idea of a topology, it is not obvious why that particular set of axioms is chosen rather

[1]Two classic examples are Bourbaki (1966) and Kelly (1970).

than any other, and the underlying motivation only slowly becomes clear. For this reason, the particular introduction to general topology given in Section 1.4 is aimed at motivating the axioms for topology by starting with the broadest structure one can conceive with respect to which the notion of a converging sequence makes sense, and then to show how this definition is narrowed to give the standard axioms for general topology.

Other texts take a somewhat different approach and motivate the axioms for topology by starting first with a *metric* space: a special type of topological space whose underlying ideas are more intuitively accessible than are those of topology in general. In addition, metric spaces play many important roles in theoretical physics in their own right; and for these reasons we shall begin with a short introduction to the theory of such spaces. But it should be emphasised that, in general, what follows cannot be regarded as a comprehensive introduction to topology, and it should be supplemented with private study. The most I can do in the limited space available is to provide a quick introduction to some of the key ideas. However, I have also included topics that I feel are of potential interest in theoretical physics but which do not appear in the standard texts on topology: a good example is the lattice structure on the set of all topologies on a given set.

## 1.2 Metric Spaces

### 1.2.1 The simple idea of convergence

A key ingredient in any topological-type structure on a set $X$ is the sense in which a point[2] $x \in X$ can be said to be 'near' to another point $y \in X$—without such a concept, the points in $X$ are totally disconnected from each other. In particular, we would like to say that an infinite sequence $(x_1, x_2, \ldots)$ of points in $X$ 'converges' to a point $x \in X$ if the elements of the sequence get arbitrarily near to $x$ in an appropriate way. We shall use the idea of the convergence of

[2]The notation $x \in X$ means that $x$ is an element of the set $X$.

sequences to develop the theory of metric spaces and, in Section 1.4, general topological spaces. As we shall see, in the latter case it is necessary to extend the discussion to include the idea of the convergence of collections of *subsets* of $X$—with this proviso, the structure of a topological space is completely reflected by the convergent collections that it admits.

A familiar example is provided by the complex numbers: the 'nearness' of one number $z_1$ to another $z_2$ is measured by the value of the modulus $|z_1 - z_2|$, and to say that the sequence $(z_1, z_2, \ldots)$ 'converges' to $z$ means that, for all real numbers $\epsilon > 0$, there exists an integer $n_0$ (which, in general, will depend on $\epsilon$) such that $n > n_0$ implies $|z_n - z| < \epsilon$; this is illustrated in Figure 1.1. Thus the disks[3] [4] [5]

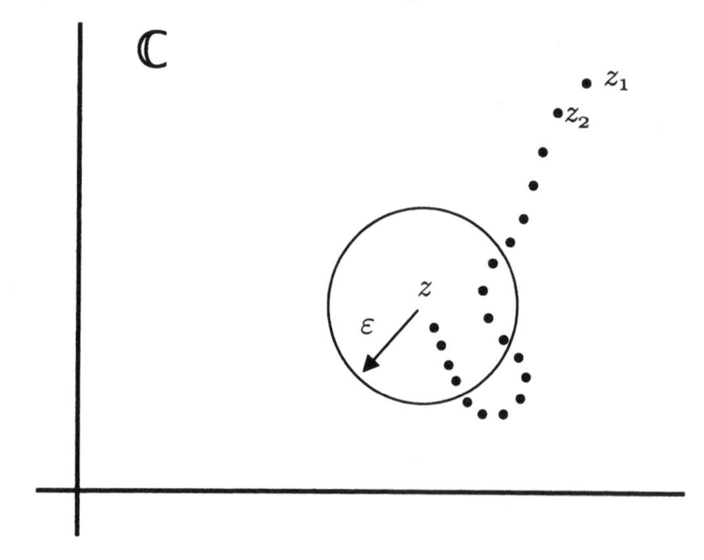


Figure 1.1: A convergent sequence of complex numbers.

$B_\epsilon(z) := \{ z' \in \mathbb{C} \mid |z - z'| < \epsilon \}$ 'trap' the sequence. That is, the

[3]An equation of the form $\alpha := \beta$ means that the quantity denoted by $\alpha$ is *defined* by the expression, $\beta$, on the right hand side.

[4]The symbol **C** denotes the complex numbers.

[5]The notation $\{x \mid P(x)\}$ means the set of all $x$ such that the proposition $P(x)$ is true.

convergence condition can be rewritten as

$$\text{“for all } \epsilon > 0 \text{ there exists } n_0 \text{ such that } n > n_0 \text{ implies } z_n \in B_\epsilon(z)\text{”} \tag{1.2.1}$$

or, in terms of the *tails* $T_n := \{ z_k \mid k > n \}$ of the sequence,[6]

$$\text{“for all } \epsilon > 0 \text{ there exists } n_0 \text{ such that } T_{n_0} \subset B_\epsilon(z)\text{”}. \tag{1.2.2}$$

This notion of convergence can be generalized at once to the space[7] $\mathbb{R}^n$ of all $n$-tuples of real numbers with the aid of the *distance function* [8]

$$d(\vec{x}, \vec{y}) := \sqrt{(\vec{x} - \vec{y}) \cdot (\vec{x} - \vec{y})} \tag{1.2.3}$$

and the associated balls

$$B_\epsilon(\vec{x}) := \{ \vec{y} \in \mathbb{R}^n \mid d(\vec{x}, \vec{y}) < \epsilon \}. \tag{1.2.4}$$

Then a sequence of points $\vec{x}_n \in \mathbb{R}^n$ is said to *converge* to $\vec{x} \in \mathbb{R}^n$ (denoted $\vec{x}_n \to \vec{x}$) if

$$\text{“for all } \epsilon > 0 \text{ there exists } n_0 \text{ such that } n > n_0 \text{ implies } \vec{x}_n \in B_\epsilon(\vec{x})\text{”}. \tag{1.2.5}$$

### 1.2.2 The idea of a metric space

The concept of a distance function can be generalised to an arbitrary set $X$ by extracting the crucial properties *vis-a-vis* convergence of the Euclidean distance $d(\vec{x}, \vec{y})$ defined in Eq. (1.2.3). This gives rise to the following definitions.

---

[6]The notation $A \subset B$ means that $A$ is a subset of $B$. This does not exclude the possibility that $A = B$. If $A \subset B$ and $A \neq B$ then $A$ is said to be a *proper* subset of $B$.

[7]The symbol $\mathbb{R}$ denotes the real numbers.

[8]The symbol $\vec{x}$ denotes the vector with components $(x_1, x_2, \ldots, x_n)$; $\vec{a} \cdot \vec{b}$ denotes the usual 'dot' product $\vec{a} \cdot \vec{b} := \sum_{i=1}^n a_i b_i$.

**Definition 1.1**

1. A *metric* on a set $X$ is a map[9] [10] $d : X \times X \to \mathbb{R}$ that satisfies the three conditions

$$
\begin{aligned}
d(x, y) &= d(y, x) && (1.2.6)\\
d(x, y) &\geq 0, \text{ and } = 0 \text{ if, and only if, } x = y && (1.2.7)\\
d(x, y) &\leq d(x, z) + d(z, y) && (1.2.8)
\end{aligned}
$$

   for all $x, y, z \in X$.

   The set $X$ itself is said to be a *metric space*. Sometimes this term is applied to the pair $(X, d)$ if it is appropriate to make a reference to the specific metric function, $d$, involved.

2. If Eq. (1.2.7) is replaced by the weaker condition "$d(x, y) \geq 0$, with $d(x, x) = 0$ for all $x \in X$" (*i.e.*, there may be $x \neq y$ such that $d(x, y) = 0$) then $X$ is said to be a *pseudo-metric* space, and the function $d$ is a *pseudo-metric*.

3. As in the example of the complex numbers, *convergence* of a sequence in a metric space can be defined in terms of the tails of the sequence being trapped by the balls surrounding a point. That is, $x_n \to x$ means

$$\text{“for all } \epsilon > 0, \text{ there exists } n_0 \text{ such that } T_{n_0} \subset B_\epsilon(x)\text{”}, \quad (1.2.9)$$

   where $B_\epsilon(x) := \{\, y \in X \mid d(x, y) < \epsilon \,\}$.

**Comments**

1. Any given sequence of points in a metric space $(X, d)$ may not converge at all but, if it does, it converges to one point only [Exercise!]. In more general types of topological space, a sequence may converge to more than one point (see later).

2. It is important to know when two metrics can be regarded as being equivalent. For example, metrics $d^{(1)}(x, y)$ and $d^{(2)}(x, y)$ on a

[9] The notation $f : A \to B$ means that $f$ is a function (or map) from the set $A$ to the set $B$. We also write $a \mapsto f(a)$ to denote that the particular element $a$ in $A$ is mapped to $f(a)$ in $B$.

[10] If $A$ and $B$ are sets, $A \times B$ denotes the *Cartesian product* of $A$ with $B$. This is defined to be the set of all (ordered) pairs $(a, b)$ where $a \in A$ and $b \in B$.

set $X$ are said to be *isometric* if there exists a bijection $\iota : X \to X$ such that, for all $x, y \in X$,

$$d^{(1)}(x,y) = d^{(2)}(\iota(x), \iota(y)). \tag{1.2.10}$$

3. Of greater interest perhaps is when two metrics lead to the same set of convergent sequences, and with each sequence converging to the same point in both metrics. This motivates the following definition:

**Definition 1.2**

1. A metric $d^{(2)}$ is *stronger* than a metric $d^{(1)}$ (or $d^{(1)}$ is *weaker* than $d^{(2)}$) if

$$\text{“for all } x \in X, \text{ for all } \epsilon > 0, \text{ there exists } \epsilon' > 0 \text{ such that } B^{(2)}_{\epsilon'}(x) \subset B^{(1)}_{\epsilon}(x)\text{”}. \tag{1.2.11}$$

2. A pair of metrics are *equivalent* if each one is stronger than the other.

**Comments**

1. For any given pair of metrics on a set $X$ it is not necessarily the case that one is either stronger or weaker than the other.

2. If the metric $d^{(2)}$ is stronger than the metric $d^{(1)}$ in the sense above, then the topology associated with $d^{(2)}$ is stronger than that associated with $d^{(1)}$ in the sense that it has 'more' open sets; see Definition 1.11 for details.

3. If $d^{(2)}$ is stronger than $d^{(1)}$, then a $d^{(2)}$-convergent sequence is automatically $d^{(1)}$-convergent [Exercise!].

4. It follows that equivalent metrics admit the same set of convergent sequences.

A result of considerable importance is the converse to this. Namely, it can be shown that if two metrics induce the same set of convergent sequences (with the same limits) then they are necessarily equivalent. Some of the material needed to prove this will be introduced later. □

### 1.2.3 Examples of metric spaces

1. If a differentiable manifold $\Sigma$ is equipped with a Riemannian metric $g$, the distance between a pair of points $x, y \in \Sigma$ is defined to be

$$d(x,y) := \inf_{\gamma} \int \left(g_{ab}(\gamma(t))\dot{\gamma}^a(t)\dot{\gamma}^b(t)\right)^{\frac{1}{2}} dt \tag{1.2.12}$$

where the *infinum* is over all piece-wise differentiable curves $t \mapsto \gamma(t)$ in $\Sigma$ that pass through the points $x$ and $y$.

2. A metric can be defined on any set $X$ by

$$d(x,y) := \begin{cases} 1 & \text{if } x \neq y \\ 0 & \text{if } x = y. \end{cases} \tag{1.2.13}$$

3. Some equivalent metrics on $\mathbb{R}^n$ are

$$d(\vec{x},\vec{y}) := \sqrt{(\vec{x}-\vec{y})\cdot(\vec{x}-\vec{y})} \tag{1.2.14}$$

$$d(\vec{x},\vec{y}) := \max_{i=1\dots n} |x_i - y_i| \tag{1.2.15}$$

$$d(\vec{x},\vec{y}) := \left(\sum_{i=1}^{n} |x_i - y_i|^p\right)^{\frac{1}{p}} \qquad p \geq 1. \tag{1.2.16}$$

When $\mathbb{R}^n$ is viewed as a topological space in this way, I shall refer to it as the *euclidean* space $\mathbb{R}^n$; usually, this will also mean using the particular metric Eq. (1.2.14).

Of course, the set of $n$-tuples $\mathbb{R}^n$ also has a natural vector space structure: if it is desirable to emphasise this property, I shall refer to the *vector* space $\mathbb{R}^n$.

4. We recall that if $\mathcal{V}$ is a complex vector space, a *norm* on $\mathcal{V}$ is a real-valued map $\vec{u} \mapsto \|\vec{u}\|$ that satisfies the three conditions

$$\text{(a)} \quad \|\vec{u}+\vec{v}\| \leq \|\vec{u}\| + \|\vec{v}\| \text{ for all } \vec{u},\vec{v} \in \mathcal{V} \tag{1.2.17}$$

$$\text{(b)} \quad \|\mu\vec{v}\| = |\mu|\,\|\vec{v}\| \text{ for all } \mu \in \mathbb{C} \text{ and } \vec{v} \in \mathcal{V} \tag{1.2.18}$$

$$\text{(c)} \quad \|\vec{v}\| \geq 0 \text{ with } \|\vec{v}\| = 0 \text{ only if } \vec{v} = 0. \tag{1.2.19}$$

An example is a Hilbert space with $\| \vec{v} \| := \sqrt{\langle v, v \rangle}$. Note that Eq. (1.2.14) is an example of this construction on the vector space $\mathbb{R}^n$.

A norm gives rise to a metric on $\mathcal{V}$ defined by

$$d(\vec{u}, \vec{v}) := \| \vec{u} - \vec{v} \| \,. \tag{1.2.20}$$

A sequence of vectors that converges with respect to this metric is said to be *strongly convergent*.

5. Spaces of functions play important roles in theoretical physics, and it is a matter of considerable significance that many of these are metric spaces. For example, let $C([a,b], \mathbb{R})$ denote the space of all real-valued, continuous functions on the closed interval $[a,b] := \{ r \in \mathbb{R} \mid a \leq r \leq b \}$. A metric can be defined on $C([a,b], \mathbb{R})$ by

$$d(f,g) := \int_a^b |f(t) - g(t)| \, dt. \tag{1.2.21}$$

Another metric is

$$d(f,g) := \sup_{t \in [a,b]} |f(t) - g(t)| \tag{1.2.22}$$

as sketched in Figure 1.2. This is inequivalent to the metric in Eq. (1.2.21).

Another inequivalent metric is

$$d(f,g) := \left\{ \int_a^b |f(t) - g(t)|^2 \, dt \right\}^{\frac{1}{2}} . \tag{1.2.23}$$

6. On any set $X$, let $\ell^2(X)$ denote the set of all real-valued functions $f$ on $X$ such that

(i) $f(x) = 0$ for all but a countable set of $x \in X$;

(ii) $\sum_{x \in X} (f(x))^2$ converges.

Then a distance function can be defined on $\ell^2(X)$ by

$$d(f,g) := \sqrt{\sum_{x \in X} |f(x) - g(x)|^2}. \tag{1.2.24}$$

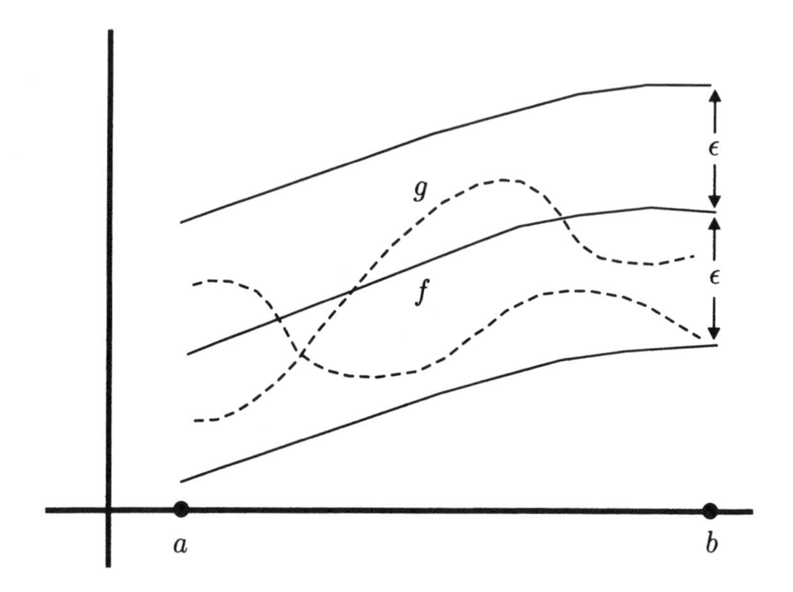


Figure 1.2: A pair of functions satisfying $d(f,g) := \sup_{t\in[a,b]} |f(t) - g(t)| < \epsilon$.

### 1.2.4 Operations on metrics

There are a number of ways in which metrics on a set $X$ may be combined to form a new metric. Some specific examples of such operations are as follows.

1. If $d_i$, $i = 1, 2, \ldots, n$ is a finite set of metrics on $X$ then

$$d(x,y) := \sum_{i=1}^{n} a_i d_i(x,y) \tag{1.2.25}$$

defines a metric on $X$ if $\{a_1, a_2, \ldots, a_n\}$ is any set of real numbers, each of which is greater than or equal to zero and such that at least one of them is non-zero.

2. If $d_1$ and $d_2$ are a pair of metrics on $X$, a new metric, called the *join* of $d_1$ and $d_2$, can be defined by, for all $x, y \in X$,

$$d_1 \vee d_2(x,y) := \max(d_1(x,y), d_2(x,y)). \tag{1.2.26}$$

One might expect to be able to use this pair of metrics to define another metric as $\min(d_1(x,y), d_2(x,y))$ but, however, this fails to satisfy the triangle inequality Eq. (1.2.8). This can be remedied by defining instead the *meet* of $d_1$ and $d_2$ to be, for all $x, y \in X$,

$$d_1 \wedge d_2(x,y) := \inf_{x=x_1 \ldots y=x_r} \sum_{k=2}^{r} \min(d_1(x_{k-1}, x_k), d_2(x_{k-1}, x_k)) \quad (1.2.27)$$

where the *infinum* is taken over all finite subsets $\{x = x_1, x_2, \ldots, x_r = y\}$ of $X$. It is interesting to note that the set of all metrics on $X$ forms a lattice under these two operations (see Section 1.3.2 for a short introduction to lattices).

3. If $d$ is any metric on $X$, define $d_b(x,y) := \min(1, d(x,y))$ for all $x, y \in X$. Then $d_b$ is a *bounded*[11] metric that can be shown to be equivalent to $d$. Thus if we are only interested in metrics up to equivalence, nothing is lost by requiring them to be bounded functions on $X \times X$.

### 1.2.5 Some topological concepts in metric spaces

In the present context, by 'topological' concepts I mean those dealing with the relations of points and subsets[12]. For example, if $A$ is a subset of the metric space $X$, then—in purely set-theoretic terms—every point in $X$ is one of just two types in relation to $A \subset X$: either (i) $x$ belongs to $A$; or (ii) it does not, in which case it belongs to the complement $A^c$ of $A$ defined as[13] $A^c := \{ x \in X \mid x \notin A \}$. However, if a metric is present on $X$ this classification can be refined to one in which any $x \in X$ belongs to one of *three* categories, defined as follows (see Figure 1.3).

[11]In general, a function $f : A \to \mathbb{R}$ is said to be *bounded* if there exists some finite real number $K > 0$ such that $|f(a)| < K$ for all $a \in A$.

[12]The word 'topological' comes from the Greek τοπος which, roughly speaking, means 'place'.

[13]The complement of a subset $A$ of a set $X$ is often written as $X - A$.

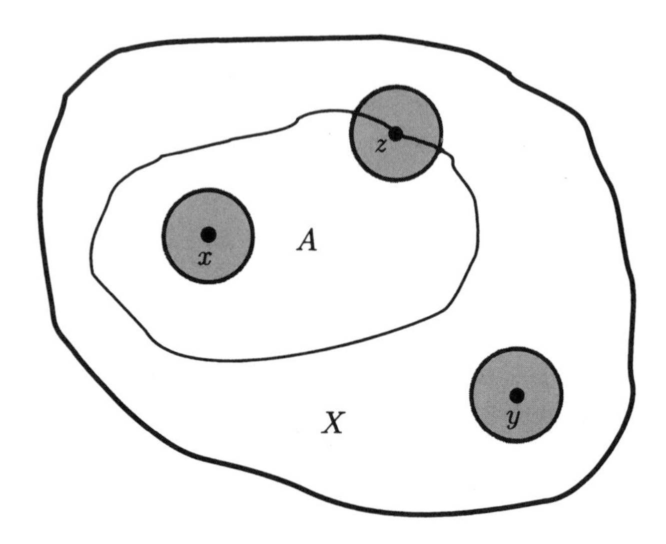


Figure 1.3: Illustration of an interior, an exterior, and a boundary point.

**Definition 1.3**

1. $x$ is an *interior* point of $A$ if there exists $\epsilon > 0$ such that the ball $B_\epsilon(x)$ around $x \in X$ has the property that $B_\epsilon(x) \subset A$.

2. $y$ is an *exterior* point of $A$ if there exists $\epsilon > 0$ such that the ball $B_\epsilon(y)$ has the property that $B_\epsilon(y) \cap A = \emptyset$.

3. $z$ is a *boundary* point of $A$ if every ball $B_\epsilon(z)$, $\epsilon > 0$, intersects both $A$ and $A^c$.

Furthermore, a point $x \in X$ is said to be a *limit* point of $A$ if $B_\epsilon(x) \cap A \neq \emptyset$ for all $\epsilon > 0$. Thus a limit point is either an interior point or a boundary point of $A$.

## Comments

1. The *interior*, *exterior*, and *boundary* of $A$ are defined to be the set of all interior, exterior and boundary points of $A$ respectively; they are denoted $\text{Int}(A)$, $\text{Ext}(A)$ and $\text{Bd}(A)$. It is easy to see that

$$\text{Int}(A) = \text{Ext}(A^c), \tag{1.2.28}$$

$$\text{Ext}(A) = \text{Int}(A^c), \tag{1.2.29}$$

$$\text{Bd}(A) = \text{Bd}(A^c), \tag{1.2.30}$$

plus

$$\text{Int}(A) \subset A, \tag{1.2.31}$$

$$A \cap \text{Ext}(A) = \emptyset. \tag{1.2.32}$$

2. A set $A$ is said to be *open* if it contains none of its boundary points. It is *closed* if it contains all its boundary points. Note that [Exercise!]

(i) a set $A$ is open if, and only if, $A = \text{Int}(A)$;

(ii) a set $A$ is open if, and only if, $A^c$ is closed.

In the example of the real line $\mathbb{R}$ with its usual metric $d(x, y) := |x - y|$, the interval $\{ x \in \mathbb{R} \mid a < x < b \}$ (for any $a < b$) is an open set; similarly $\{ x \in \mathbb{R} \mid a \leq x \leq b \}$ is an example of a closed set. On the other hand, $\{ x \in \mathbb{R} \mid a \leq x < b \}$ is neither open nor closed.

3. The collection of all open sets in any metric space is called the *topology* associated with the space. It possesses the following important properties:

- the union of an *arbitrary* collection of open sets is open;
- the intersection of any *finite* collection of open sets is open;
- The empty set $\emptyset$ and $X$ itself are both open.

These properties are of fundamental importance—indeed, a general topological space is *defined* to be a set $X$ with a family of subsets that satisfy these three properties. This will be discussed in detail later.

4. The analogous properties for closed sets are:

- the intersection of an *arbitrary* collection of closed sets is closed;

- the union of any *finite* collection of closed sets is closed;
- the empty set $\emptyset$ and $X$ itself are closed.

5. The topology associated with a metric space is determined equally by either the collection of all open sets or the collection of all closed sets. In the latter context, it is therefore significant that:

(i) $A \subset X$ is closed if and only if it contains all its limit points;

(ii) a point $x \in X$ is a limit point of a subset $A$ if and only if there exists a sequence $(x_1, x_2, \ldots)$ in $A$ that converges to $x$.

Thus a subset $A$ is closed if, and only if, the limit of every convergent sequence $(x_1, x_2, \ldots)$ of points in $A$ itself lies in $A$. It follows that the closed sets (and hence the topology) associated with a metric are uniquely determined by its collection of convergent sequences. This is the key to proving the result mentioned earlier that two metrics with the same set of convergent sequences are equivalent.

6. We will see later the precise sense in which a metric space is a special case of a general topological space. Thus the topological differences between, for example, a 2-sphere, a 2-torus and a 1264-sphere are coded entirely in their respective distance functions, all of which could be considered to be defined on some common abstract set $X$ with the cardinality of the continuum. □

## 1.3 Partially Ordered Sets and Lattices

### 1.3.1 Partially ordered sets

In developing the general theory of topology, it is useful to emphasise certain algebraic properties that arise naturally in this context. The relevant concepts are 'partially ordered set' and 'lattice', both of which play an important role in many branches of mathematics.[14]

[14] A useful reference with applications to quantum theory is Beltrametti & Cassinelli (1981).

**Definition 1.4**

1. A *relation* $R$ on a set $X$ is a subset of $X \times X$; and $x \in X$ is said to be *R-related* to $y \in X$ (denoted $xRy$) if the pair $(x, y) \in R \subset X \times X$. Note that a function $f : X \to X$ defines a relation $\{(x, f(x)) \mid x \in X\}$; however, there are many relations that are not derived from functions.

2. A *partially ordered set* (or *poset*) is a set $X$ with a relation $\preceq$ on $X$ that is:

   (P1) *Reflexive*: for all $x \in X$, $x \preceq x$.

   (P2) *Antisymmetric*: for all $x, y \in X$, if $x \preceq y$ and $y \preceq x$ then $x = y$.

   (P3) *Transitive*: for all $x, y, z \in X$, if $x \preceq y$ and $y \preceq z$ then $x \preceq z$.

   A *pre-order* on a set $X$ is a relation that is reflexive and transitive. Thus a poset is a pre-ordered set whose associated relation is also antisymmetric.

   The notation $x \prec y$ will be used if it is necessary to emphasise that $x \preceq y$ but $x \neq y$. Note that any particular pair of elements $x, y \in X$ may not be related either way. However, if it is true that for any $x, y \in X$ either $x \preceq y$ or $y \preceq x$, then $X$ is said to be *totally* ordered.

3. An element $y$ in a poset $X$ *covers*[15] another element $x$ if $x \prec y$ and there is no $z \in X$ such that $x \prec z \prec y$. This is denoted diagramatically by

   It is clear that a finite, partially ordered set is determined uniquely by its diagram of covering elements.

[15] It is perfectly possible for an element to cover more than one element.

4. For later use we recall also the definition of an *equivalence relation* on a set $X$. This is a relation $R$ that is:

   (E1) *Reflexive*: for all $x \in X$, $xRx$.

   (E2) *Symmetric*: for all $x, y \in X$, $xRy$ implies $yRx$.

   (E3) *Transitive*: for all $x, y, z \in X$, $xRy$ and $yRz$ implies $xRz$.

It should be noted that any equivalence relation $R$ on a set $X$ partitions $X$ into disjoint equivalence classes in which all the elements in any class are equivalent to each other. The set of all such equivalence classes is denoted $X/R$. An important example in theoretical physics is the set of gauge orbits of the action of a gauge group on the space of vector potentials in a Yang-Mills theory.

**Examples**

1. The real numbers $\mathbb{R}$ are totally ordered with respect to the usual ordering $\leq$, in which $a \leq b$ means that $a \in \mathbb{R}$ is less than, or equal to, $b \in \mathbb{R}$. Note that no $r \in \mathbb{R}$ possesses a cover since, given any pair of real numbers, there always exists a third one that lies between them.

On the other hand, the integers $\mathbf{Z}$ are also totally ordered with respect to $\leq$, but each $n \in \mathbf{Z}$ *does* possess a unique cover, namely $n+1$.

2. A pre-order $\preceq$ can be defined on the set Metric$(X)$ of all metric functions on a set $X$ by saying that $d^{(1)} \preceq d^{(2)}$ if the open balls for the two metrics satisfy Eq. (1.2.11). This means that the topology associated with $d^{(2)}$ is stronger than that associated with $d^{(1)}$.

As we shall see later, topologies in general can be partially ordered by the relation of one being stronger than the other. However, the pre-order $\preceq$ is *not* a partial ordering on the set Metric$(X)$ since it is not antisymmetric: $d^{(1)} \preceq d^{(2)}$ and $d^{(2)} \preceq d^{(1)}$ do not imply that $d^{(1)} = d^{(2)}$ but only that the two metrics are equivalent (that is, they admit the same set of convergent sequences).

3. If $X$ is any set, the set of all subsets of $X$ is denoted P$(X)$ and is known as the *power set* of $X$. Thus $A \subset X$ if, and only if,

$A \in \mathrm{P}(X)$. In the general theory of topological convergence (to be developed later) the central idea is to associate with each $x \in X$ a collection $\mathcal{N}(x)$ of subsets of $X$ (the 'neighbourhoods' of the point $x$) that determine whether or not a sequence converges to $x$. Thus $\mathcal{N}(x) \subset \mathrm{P}(X)$; or, equivalently, $\mathcal{N}(x) \in \mathrm{P}(\mathrm{P}(X))$. Similarly, the collection $\mathcal{N} := \{\mathcal{N}(x) \mid x \in X\}$ can be regarded as a subset of $\mathrm{P}(\mathrm{P}(X))$ or as an element of $\mathrm{P}(\mathrm{P}(\mathrm{P}(X)))$.

The set $\mathrm{P}(X)$ has a natural partial ordering defined by

$$A \preceq B \text{ means } A \subset B \tag{1.3.1}$$

where $A, B \subset X$. In this example, a cover of a subset $A$ is any subset of $X$ that is obtained by adding a single point to $A$.

The simplest non-trivial example is the partial-ordering diagram for the two-element set $X = \{a, b\}$ with $P(X) = \{\emptyset, X, \{a\}, \{b\}\}$:

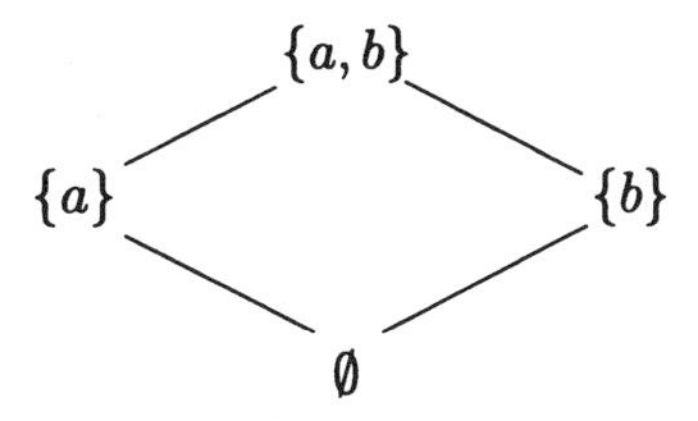


while the diagram for $X = \{a, b, c\}$ is

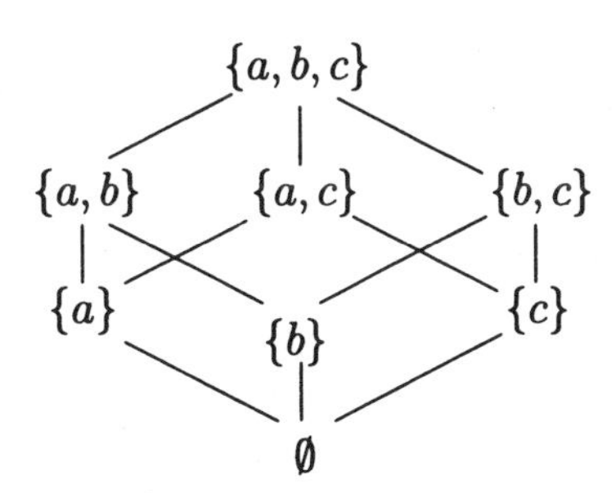

4. Partially ordered sets play an important role in classical general relativity. Specifically, let $\mathcal{M}$ be a spacetime manifold equipped with a Lorentzian metric. Then if $q, p \in \mathcal{M}$, define $q \preceq p$ if $p$ lies in the causal future of $q$: *i.e.*, $q$ can be joined to $p$ by a path whose tangent vector is everywhere timelike or null (Kronheimer & Penrose 1967). This is a partial ordering, and—rather remarkably—the entire metric (up to an overall conformal factor) can be recovered from this ordering (Hawking, King & McCarthy 1976, Malament 1977). This feature has been behind a variety of suggestions that spacetime should be regarded as a discrete set but still with a causal structure/partial ordering[16].

### 1.3.2 Lattices

One of the central branches of mathematics in which a poset structure occurs naturally is propositional logic where, given any two propositions $a$ and $b$, the relation $a \leq b$ is defined to mean that $a$ logically *implies* $b$; *i.e.*, if $a$ is true then $b$ is necessarily true. However, in this particular example there is important extra structure given by the propositional functions '*and*' and '*or*'; more precisely, given any pair of propositions $a$ and $b$ we can form the propositions '$a$ *and* $b$' and '$a$ *or* $b$', usually denoted $a \wedge b$ and $a \vee b$ respectively. Analogues of these and related operations arise in many different situations, and are captured in the following definitions:

**Definition 1.5**

1. In any poset $\mathcal{P}$, a *meet* (or *greatest lower bound*) of $a, b \in \mathcal{P}$ is an element $a \wedge b \in \mathcal{P}$ such that:

   (a) $a \wedge b$ is a *lower bound* of $a$ and $b$: thus $a \wedge b \preceq a$ and $a \wedge b \preceq b$;

   (b) $a \wedge b$ is the greatest such lower bound: *i.e.*, if there exists $c \in \mathcal{P}$ such that $c \preceq a$ and $c \preceq b$ then $c \preceq a \wedge b$.

[16]Rafael Sorkin has been one of the most articulate exponents of this idea; for example, see Sorkin (1991)

2. A *join* (or *least upper bound*) of $a, b \in \mathcal{P}$ is an element $a \vee b \in \mathcal{P}$ such that:

   (a) $a \vee b$ is an *upper bound* of $a$ and $b$: thus $a \preceq a \vee b$ and $b \preceq a \vee b$;

   (b) $a \vee b$ is the least such upper bound: *i.e.*, if there exists $c \in \mathcal{P}$ such that $a \preceq c$ and $b \preceq c$ then $a \vee b \preceq c$.

   Note that, if it exists, a join or meet is necessarily unique [Exercise!].

3. A *lattice* is a poset $\mathcal{L}$ in which every pair of elements possesses a join and a meet. A *unit* element in a lattice $\mathcal{L}$ is an element 1 such that, for all $a \in \mathcal{L}$, $a \preceq 1$. A *null* element in a lattice $\mathcal{L}$ is an element 0 such that, for all $a \in \mathcal{L}$, $0 \preceq a$.

4. The lattice is *complete* if a greatest lower bound and a least upper bound exist for *every* subset[17] $S$ of $\mathcal{L}$ (all that is guaranteed by the definition of a lattice is that these bounds will exist for all *finite* subsets of $\mathcal{L}$). If they exist, these bounds are denoted $\bigwedge S$ and $\bigvee S$ respectively.

5. A lattice is *distributive* if, for all $a, b, c \in \mathcal{L}$, we have

$$a \wedge (b \vee c) = (a \wedge b) \vee (a \wedge c) \tag{1.3.2}$$

   and

$$a \vee (b \wedge c) = (a \vee b) \wedge (a \vee c). \tag{1.3.3}$$

6. A map $a \mapsto a'$ of a poset $\mathcal{P}$ is an *orthocomplementation* if, for all $a, b \in \mathcal{P}$,

$$\text{(i) } (a')' = a \tag{1.3.4}$$

$$\text{(ii) } a \preceq b \text{ implies } b' \preceq a' \tag{1.3.5}$$

$$\text{(iii) } a \vee a' = 1 \text{ and } a \wedge a' = 0. \tag{1.3.6}$$

   A poset with an orthocomplementation operation is said to be *orthocomplemented*.

[17] By a *least upper bound* of a subset $S$ of $\mathcal{L}$ is meant an element $u \in \mathcal{L}$ such that (i) for all $a \in S$, $a \leq u$; and (ii) if $v \in \mathcal{L}$ is any other element such that $a \leq v$ for all $a \in S$ then $u \leq v$. The definition of a greater lower bound is analogous.

7. A *Boolean algebra* is an orthocomplemented, distributive lattice.

**Comments**

1. All the lattices we shall be considering have both a unit element and a null element.

2. $a \preceq b$ if and only if $a \wedge b = a$ if and only if $a \vee b = b$.

3. For all $a \in \mathcal{L}$, $1 \wedge a = a$. Thus $\mathcal{L}$ is a semigroup[18] with respect to the $\wedge$-operation with 1 as the unit element; however, it is not a group since no element other than 1 has an inverse.

Similarly, $0 \vee a = a$, and hence $\mathcal{L}$ is also a semigroup with respect to the $\vee$-operation with 0 as the unit element.

In addition, for all $a \in \mathcal{L}$,

$$1 \vee a = 1, \text{ and } 0 \wedge a = 0. \tag{1.3.7}$$

Thus 1 and 0 are *absorptive* elements for the $\vee$-semigroup and the $\wedge$-semigroup respectively.

4. If $\mathcal{L}$ is distributive then any complement $a'$ of an element $a \in \mathcal{L}$ is unique.

□

**Examples**

1. A simple example of a Boolean algebra is given by the following diagram in which $a \wedge b = 0$ and $a \vee b = 1$:

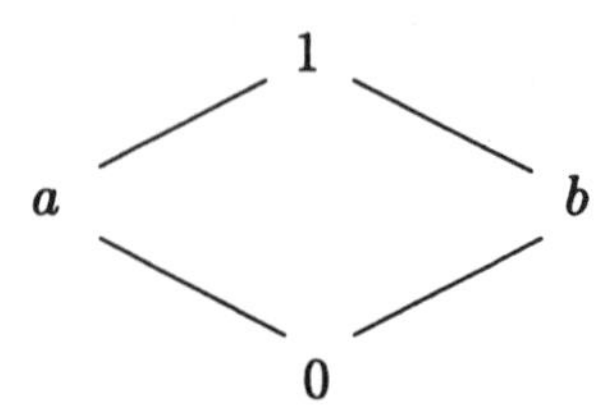


[18] A semigroup is a set equipped with an associative combination law and a unit element. However, unlike the case of a group, inverse elements need not exist.

2. Two non-distributive lattices with 5 elements:

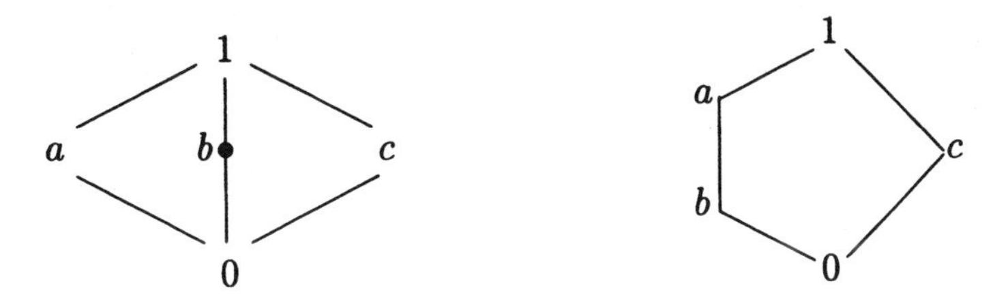


3. The set P($X$) of all subsets of a set $X$ is a Boolean algebra with

$$
\begin{aligned}
A \wedge B &:= A \cap B && (1.3.8)\\
A \vee B &:= A \cup B && (1.3.9)
\end{aligned}
$$

and $A \preceq B$ defined to mean $A \subset B$. More generally, if $A_i$, $i \in I$ is any family of subsets of $X$, the meet and join of the family is defined to be the intersection and union respectively of the sets in the family.

The unit and null elements are

$$
\begin{aligned}
1 &:= X && (1.3.10)\\
0 &:= \emptyset. && (1.3.11)
\end{aligned}
$$

The lattice theory complement $A'$ of $A \subset X$ is defined to be the set-theoretic complement $A^c$. Clearly this lattice is complete.

4. The set of all metrics on a set $X$ can be partially ordered by defining $d^{(1)} \preceq d^{(2)}$ if $d^{(1)}(x,y) \leq d^{(2)}(x,y)$ for all $x, y \in X$ (this should not be confused with the pre-order relation given earlier in Definition 1.2). It becomes a lattice under the join and meet operations defined in Eqs. (1.2.26–1.2.27). Note, however, there is no null or unit element since, given any metric $d(x,y)$, a larger (resp. smaller) metric can always be constructed by multiplying $d(x,y)$ by a positive real number that is greater than (resp. less than) 1.

5. If $V$ is a vector space, the set of all linear subspaces of $V$ is a lattice with

$$W_1 \wedge W_2 := W_1 \cap W_2 \tag{1.3.12}$$

and with $W_1 \vee W_2$ defined to be the smallest subspace that contains the pair of subspaces $W_1$ and $W_2$.

If $V$ is a Hilbert space, the lattice of *closed* linear subspaces[19] is complemented, where the complement of a subspace $W$ is defined to be its orthogonal complement with respect to the Hilbert space inner product $<,>$:

$$W' := W_{\perp} = \{\, v \in V \mid \text{for all } w \in W,\ <v, w>= 0 \,\}. \tag{1.3.13}$$

Correspondingly, the set of all hermitian projection operators on $V$ also forms a complemented lattice.

This particular lattice has been used extensively in investigations into the axiomatic foundations of general quantum theory. It differs strikingly from the analogous lattice of propositions in classical mechanics in that it is *not* distributive. This is because the basic type of yes-no question that can be asked in classical physics is whether the point in state space that represents the state of the system does, or does not, lie in any particular subset. Thus the propositional lattice of classical physics is essentially the lattice of subsets of state space, which is automatically distributive.[20] This distinction between the quantum and classical lattices has given rise to the interesting subject of 'quantum logic' (Beltrametti & Cassinelli 1981, Isham 1995). □

A lattice $\mathcal{L}$ satisfies several very important algebraic relations:

(L1) *Idempotency*: $a \vee a = a$ and $a \wedge a = a$, for all $a \in \mathcal{L}$.

(L2) *Commutativity*: $a \vee b = b \vee a$ and $a \wedge b = b \wedge a$, for all $a, b \in \mathcal{L}$.

(L3) *Associativity*: $(a \vee b) \vee c = a \vee (b \vee c)$ and $(a \wedge b) \wedge c = a \wedge (b \wedge c)$, for all $a, b, c \in \mathcal{L}$.

In addition, any lattice satisfies the *absorptive* laws:

---

[19] If $V$ has an infinite dimension, the disjunction $W_1 \vee W_2$ of two closed subspaces is defined to be the *closure* of the smallest subspace $[W_1, W_2]$ that contains both $W_1$ and $W_2$; the subspace $[W_1, W_2]$ itself may not be closed.

[20] More precisely, the lattice is usually chosen to be the set of all subsets that are *measurable* with respect to some given measure structure.

(L4) $a \wedge (a \vee b) = a$ and $a \vee (a \wedge b) = a$ for all $a, b \in \mathcal{L}$.

Conversely, there is the important theorem:

**Theorem 1.1** *Any non-empty set $\mathcal{L}$ that is equipped with binary operations $\wedge$ and $\vee$ that satisfy the conditions (L1)–(L4) can be given a partial ordering by defining*

$$a \preceq b \text{ if } a = a \wedge b. \tag{1.3.14}$$

*The resulting structure is a lattice in which the meet and join operations are $a \wedge b$ and $a \vee b$ respectively.*

**Proof**

[Exercise!] **QED**

# 1.4 General Topology

## 1.4.1 An example of non-metric convergence

As mentioned earlier, the general theory of topology may be approached in several different ways, as reflected in the variety of styles to be found in the many textbooks that are available on the subject. We shall study the theory of general topological spaces in terms of the convergence of sequences and generalizations thereof.[21]

In a non-metric space $X$, it is no longer possible to define 'nearness' using a real number. Instead we attempt to trap the tails of a sequence with subsets of $X$ that can serve as some sort of analogue of the balls $B_\epsilon(x)$ in a metric space. A structure of this type will consist of, for each $x \in X$, a collection $\mathcal{N}(x)$ of subsets of $X$ (the 'neighbourhoods' of $x$) with convergence being defined purely in terms of these subsets.

[21] A selection of particularly useful references in this context is Bourbaki (1966), Császár (1978), Dugundji (1996), and Kelly (1970). An excellent problem-oriented introduction to set theory and topology is Lipschutz (1965).

Specifically, a sequence $(x_1, x_2, \ldots)$ in $X$ is said to *converge* to $x$ with respect to $\mathcal{N}(x)$ (denoted $x_n \xrightarrow{\mathcal{N}(x)} x$) if

$$\text{"for all } N \in \mathcal{N}(x) \text{ there exists } n_0 \text{ such that } n > n_0 \text{ implies } x_n \in N\text{"} \tag{1.4.1}$$

or, equivalently, in terms of the tails $T_n := \{x_k \mid k > n\}$ of the sequence:

$$\text{"for all } N \in \mathcal{N}(x) \text{ there exists } n_0 \text{ such that } T_{n_0} \subset N.\text{"} \tag{1.4.2}$$

Note that this does not rule out the possibility that a sequence may converge to more than one point.

At this point, it might be useful to give a well-known example of convergence that is not associated with any metric or pseudo-metric structure. Specifically, let $\mathcal{F}([a,b],\mathbb{R})$ denote the set of all real-valued functions on the closed interval $[a,b] \subset \mathbb{R}$. A sequence of functions $f_n$ is said to converge *pointwise* to a function $f$ if, for all $t \in [a,b]$, the sequence of real numbers $f_n(t)$ converges to the real number $f(t)$ in the usual way. That is[22]

$$\text{"for all } t \in [a,b], \text{ for all } \epsilon > 0, \exists\, n_0(\epsilon, t) \text{ such that } n > n_0 \text{ implies} |f_n(t) - f(t)| < \epsilon.\text{"} \tag{1.4.3}$$

An appropriate family of neighbourhoods of $f \in \mathcal{F}([a,b],\mathbb{R})$ is the collection of all finite intersections of sets of the form

$$N_{t,\epsilon}(f) := \{\, g \in \mathcal{F}([a,b],\mathbb{R}) \mid |g(t) - f(t)| < \epsilon \,\} \tag{1.4.4}$$

where $t$ is any real number in the closed interval $[a,b]$ and $\epsilon$ is any positive number; see Figure 1.4 for an illustration of a typical neighbourhood $N_{t,\epsilon}(f)$.

If a function $g$ belongs to a finite intersection of $n$ sets of this type, there is a set of points $t_1, t_2, \ldots, t_n$, and positive real numbers $\epsilon_1, \epsilon_2, \ldots, \epsilon_n$ such that the value of $g$ at these points is constrained to lie in the open intervals $(-\epsilon_1, \epsilon_1)$, $(-\epsilon_2, \epsilon_2)$, ..., $(-\epsilon_n, \epsilon_n)$ respectively.

It is interesting to note that this is the type of restriction imposed by Feynman in his analysis of finite-time approximations to a path

[22]The symbol $\exists$ is short for "there exists".

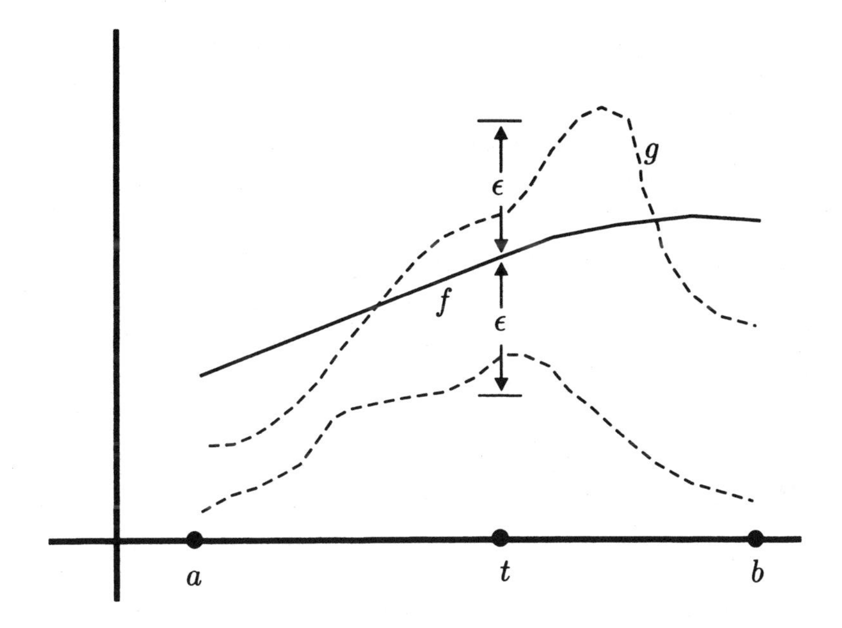


Figure 1.4: A neighbourhood in function space.

integral. In this case, $g$ could represent the position of a particle that is constrained to lie in the intervals $(-\epsilon_1, \epsilon_1)$, $(-\epsilon_2, \epsilon_2)$, ..., $(-\epsilon_n, \epsilon_n)$ at times $t_1, t_2, \ldots, t_n$ respectively.

### 1.4.2 The idea of a neighbourhood space

Two crucial questions that must be addressed are:

1. What properties should be possessed by the collections $\mathcal{N}(x)$, $x \in X$, of subsets of $X$ in order to give a notion of convergence that accords with our intuition of what this should mean?

2. When do two different families of neighbourhoods lead to the same sets of convergent sequences?

For example, in regard to the first question, a minimal requirement would seem to be that the constant sequence $x_n := x$, $n = 1, 2, \ldots$, should always converge to $x$. A necessary and sufficient condition for this is that each neighbourhood of $x$—*i.e.*, each $N \in \mathcal{N}(x)$—must contain the point $x$: a property, arguably, that is implicit in the use of the word 'neighbourhood' in the first place. Note that, in particular, this condition implies that the empty set $\emptyset$ is not a member of $\mathcal{N}(x)$ for any $x \in X$.

We could try to consider other *prima facie* requirements on these neighbourhoods, but it turns out to be more useful to start with the second question, in which case the first important concept is the following:

**Definition 1.6**

1. If $\alpha, \beta \subset \mathrm{P}(X)$, the collection $\beta$ of subsets of $X$ is *finer* than the collection $\alpha$ if, for each $A \in \alpha$, there exists a subset $B \in \beta$ such that $B \subset A$. This will be denoted by $\alpha \vdash \beta$. Equivalently, we say that the collection $\alpha$ is *coarser* than the family $\beta$.

2. Two families $\mathcal{N}^{(2)}(x)$ and $\mathcal{N}^{(1)}(x)$ of neighbourhoods of a point $x \in X$ are *equivalent* (denoted $\mathcal{N}^{(2)}(x) \cong \mathcal{N}^{(1)}(x)$) if each is finer than the other.

**Comments**

1. If $\alpha \subset \beta$ then it is trivial that $\beta$ is finer than $\alpha$.

2. The definition Eq. (1.4.2) of convergence with respect to $\mathcal{N}(x)$ is equivalent to the statement that the set $T := \{T_1, T_2, \ldots\}$ of tails of the sequence is finer than the family $\mathcal{N}(x)$, *i.e.*, $\mathcal{N}(x) \vdash T$.

3. If $\alpha \vdash \beta$ and $\beta \vdash \gamma$ then $\alpha \vdash \gamma$; thus the $\vdash$ relation is *transitive*.

In particular, if $\mathcal{N}^{(2)}(x)$ is finer than $\mathcal{N}^{(1)}(x)$ (*i.e.*, $\mathcal{N}^{(1)}(x) \vdash \mathcal{N}^{(2)}(x)$) then any sequence $(x_1, x_2, \ldots)$ that converges to $x$ with respect to $\mathcal{N}^{(2)}(x)$—so that $\mathcal{N}^{(2)}(x) \vdash T$ where $T$ denotes the set of tails of the sequence—necessarily converges with respect to $\mathcal{N}^{(1)}(x)$.

4. If a metric $d^{(2)}$ on a space $X$ is stronger than another metric $d^{(1)}$ in the sense of Definition 1.2, then $\mathcal{N}^{(1)}(x) \vdash \mathcal{N}^{(2)}(x)$ for all $x \in X$.

Here, the set of neighbourhoods of a point $x \in X$ in a metric space is defined to be the set of all open sets that contain $x$.

5. The relation $\mathcal{N}^{(2)}(x) \cong \mathcal{N}^{(1)}(x)$ defined above is an equivalence relation on subsets of $\mathrm{P}(X)$. That is:

$$\mathcal{N}(x) \cong \mathcal{N}(x) \tag{1.4.5}$$

$$\mathcal{N}^{(1)}(x) \cong \mathcal{N}^{(2)}(x) \text{ implies } \mathcal{N}^{(2)}(x) \cong \mathcal{N}^{(1)}(x) \tag{1.4.6}$$

$$\mathcal{N}^{(1)}(x) \cong \mathcal{N}^{(2)}(x) \text{ and } \mathcal{N}^{(2)}(x) \cong \mathcal{N}^{(3)}(x) \text{ implies } \mathcal{N}^{(1)}(x) \cong \mathcal{N}^{(3)}(x). \tag{1.4.7}$$

6. Two equivalent sets of neighbourhoods of $x \in X$ produce the same collection of sequences that converge to $x$.

7. The balls of two metrics that are equivalent in the sense of Definition 1.2 produce equivalent sets of neighbourhoods, where a neighbourhood of a point $x$ is defined to be any set that contains an open ball that contains $x$. □

As far as convergence of sequences is concerned, the remark in Comment 6 above implies that neighbourhoods are of interest only up to equivalence. However, the situation is not totally dissimilar to that in a gauge theory: in particular, it is useful to find the analogue of a natural 'gauge choice', *i.e.*, a collection of conditions on the elements of $\mathcal{N}(x)$ that select a unique representative from the set of equivalent collections of neighbourhoods.

In this context, the first observation is that the convergence properties of the family of neighbourhoods $\mathcal{N}(x)$ is not affected if we append to $\mathcal{N}(x)$ any subset of $X$ that is a superset of an element of this family. In any lattice $\mathcal{L}$, a subset $U$ of $\mathcal{L}$ is said to be an *upper set* if $a \in U$ implies that $b \in U$ for all $b \in \mathcal{L}$ satisfying $a \preceq b$. Thus our claim is that there is no loss of generality in assuming that each $\mathcal{N}(x)$ is an upper set in the lattice $P(X)$ of subsets of $X$.

Furthermore, it is easy to see that if $\alpha \vdash \beta$, (*i.e.*, $\alpha$ is coarser than $\beta$) and $\beta$ is upper, then $\alpha \subset \beta$. Therefore, if $\alpha$ and $\beta$ are both upper it follows that $\alpha \cong \beta$ if and only if $\alpha = \beta$. Thus each equivalence class contains precisely *one* upper family.

The second observation stems from the following lemma.

**Lemma** Given any family of subsets $\mathcal{N}(x)$, define $\mathcal{N}'(x)$ to be the collection of all finite intersections of sets belonging to $\mathcal{N}(x)$. Then a sequence in $X$ converges to $x$ with respect to $\mathcal{N}(x)$ if and only if it converges with respect to $\mathcal{N}'(x)$.

**Proof**

Clearly $\mathcal{N}'(x)$ is finer than $\mathcal{N}(x)$ (since $\mathcal{N}(x) \subset \mathcal{N}'(x)$) and hence if $(x_1, x_2, \ldots)$ converges to $x$ with respect to $\mathcal{N}'(x)$ it also converges with respect to $\mathcal{N}(x)$.

Conversely, if $x_n \xrightarrow{\mathcal{N}(x)} x$ then, for any finite collection $A_1, A_2, \ldots, A_m \in \mathcal{N}(x)$, there exists $n_1, n_2, \ldots, n_m$ such that

$$\begin{aligned} n > n_1 &\quad \text{implies} \quad x_n \in A_1 \\ n > n_2 &\quad \text{implies} \quad x_n \in A_2 \\ &\quad \vdots \\ n > n_m &\quad \text{implies} \quad x_n \in A_m. \end{aligned} \tag{1.4.8}$$

Thus $n > \max(n_1, n_2, \ldots, n_m)$ implies that $x_n \in A_1 \cap A_2 \cap \ldots \cap A_m$, and so $x_n \xrightarrow{\mathcal{N}'(x)} x$. **QED**

It follows from the above that, as far as the convergence of sequences is concerned, there is no loss of generality in choosing $\mathcal{N}(x)$ to be algebraically closed under the formation of finite intersections of its members. Thus, from this perspective, there is no loss in generality in requiring each family of sets $\mathcal{N}(x)$, $x \in X$, to be

(i) algebraically closed under finite intersections; (1.4.9)

(ii) an upper family. (1.4.10)

The two conditions Eqs. (1.4.9–1.4.10) constitute our 'gauge choice' for the elements in $\mathcal{N}(x)$. This is captured in the following important definition (which also includes the fact that $\emptyset \notin \mathcal{N}(x)$ for all $x \in X$).

**Definition 1.7**

A *filter* $\mathcal{F}$ on $X$ is a family of subsets of $X$ such that[23]

(a) $\emptyset \notin \mathcal{F}$;

(b) $\mathcal{F}$ is algebraically closed under finite intersections;

(c) $\mathcal{F}$ is an upper family.

It follows from the discussion above that our final form of the idea of convergence can be cast in terms of filters on $X$:

**Definition 1.8**

1. A *neighbourhood structure* $\mathcal{N}$ on a set $X$ is an assignment to each $x \in X$ of a filter $\mathcal{N}(x)$ on $X$ all of whose elements contain the point $x$. The pair $(X, \mathcal{N})$ (or simply $X$, if $\mathcal{N}$ is understood) is called a *neighbourhood space* (Császár 1978); the filter $\mathcal{N}(x)$ is called the *neighbourhood filter* of the point $x$ in $X$.

2. A sequence $(x_1, x_2, \ldots)$ *converges* to $x$ with respect to $\mathcal{N}$ if

$$\text{“for all } N \in \mathcal{N}(x), \text{ there exists } n_0 \text{ such that } n > n_0 \text{ implies } x_n \in N\text{”}. \tag{1.4.11}$$

3. A *filter base* $\mathcal{B}$ is a family of non-empty subsets of $X$ such that if $A, B \in \mathcal{B}$ then there exists $C \in \mathcal{B}$ such that $C \subset A \cap B$.

**Comments**

1. The definition above is about the most general notion of convergence of sequences that one could conceive; it forms the foundation for a variety of special structures in which the filters $\mathcal{N}(x)$ are restricted in some way. As we shall see shortly, a ‘topology’ is one such example.

2. Every filter is a filter base.

---

[23]More formally, this says that $\mathcal{F}$ is a ‘proper dual ideal’ in the lattice $P(X)$.

3. If $\mathcal{B}$ is a filter base then $\uparrow(\mathcal{B}) := \{ B \subset X \mid \exists A \in \mathcal{B}$ such that $A \subset B \}$ is a filter. This filter is said to be *generated* by $\mathcal{B}$, and the filter base $\mathcal{B}$ is said to be a *base* for this filter.

In practice, it is often convenient to deal with filter bases rather than the (frequently much larger) filters that they generate. Nothing is lost in doing so since a sequence converges with respect to a filter if and only if it converges with respect to any associated filter base (convergence with respect to a filter base is defined in the obvious way).

4. If $\mathcal{B}$ is a filter base, and if $\alpha \subset P(X)$ is such that $\alpha \cong \mathcal{B}$, then $\alpha$ is also a filter base.

5. A non-empty collection $\mathcal{B}$ of subsets of $X$ is a base for a specific filter $\mathcal{F}$ on $X$ if and only if

(i) $\mathcal{B} \subset \mathcal{F}$;

(ii) If $A \in \mathcal{F}$, there exists $B \in \mathcal{B}$ such that $B \subset A$.

□

We note that (i) the collection $T$ of tails of a sequence $(x_1, x_2, \ldots)$ is a filter base [Exercise!]; and (ii) the condition for convergence of this sequence can be rewritten as "$(x_1, x_2, \ldots)$ converges to $x$ with respect to $\mathcal{N}$ if and only if the filter base $T$ is finer than the filter $\mathcal{N}(x)$".

The significance of this version of the idea of convergence is that it admits a very important generalisation to an arbitrary filter base $\mathcal{B}$. This is contained in the following definition.

**Definition 1.9**

A filter base $\mathcal{B}$ *converges* to $x \in X$ if $\mathcal{B}$ is finer than $\mathcal{N}(x)$.

Note that this is true if and only if $\mathcal{B}$ is finer than any filter base $\mathcal{B}(x)$ for $\mathcal{N}(x)$. This is often useful in practice. Note also that collections of neighbourhoods $\mathcal{N}^{(1)}$ and $\mathcal{N}^{(2)}$ that are equivalent admit the same set of convergent filters.

The extension of the idea of convergence from sequences to filter bases is of considerable importance in the discussion of topological spaces that are not metric spaces. We shall return to this topic in Section 1.4.7.

**Examples**

1. In a metric space, the set of balls $B_\epsilon(x)$, $\epsilon > 0$, is a filter base for the filter $\mathcal{N}(x)$ of all neighbourhoods of $x$.

2. On the space of functions $\mathcal{F}([a,b],\mathbb{R})$, a collection of neighbourhoods of a particular function $f$ was defined in Eq. (1.4.4) as $N_{t,\epsilon}(f) := \{g \mid |g(t) - f(t)| < \epsilon\}$. These form what is known as a filter *subbase*: that is, the set of all finite intersections of sets of this type forms a filter base. □

**Comments**

1. Many of the topological concepts introduced in the context of metric spaces possess precise analogues in a neighbourhood space.

For example, an *interior point* of a set $A \subset X$ is any point $x$ such that there exists $N \in \mathcal{N}(x)$ with $N \subset A$. The definitions of *exterior* point, *boundary* point and *limit* point generalise in a similar way. It is important to note that none of these concepts change if the filter $\mathcal{N}(x)$ is replaced by any filter base that is equivalent to it.

2. Similarly, the definitions of *open* and *closed* sets are as before. And, as in the case of metric spaces, $A$ is open (i) if and only if $A^c$ is closed; and (ii) if and only if $A = \text{Int}(A)$.

In particular, a set $A$ is open if, and only if, for all $x \in A$, there exists $N \in \mathcal{N}(x)$ such that $N \subset A$.

It can be shown that, as in the case of a metric space: (i) $\emptyset$ and $X$ are open and closed; (ii) a finite intersection of open sets is an open set, as is an arbitrary union of open sets; and (iii) an arbitrary intersection of closed sets is a closed sets, as is a finite union of closed sets. [Exercise!]

3. It is also the case that $A$ is closed if and only if it contains all its limit points. □

### 1.4.3 Topological spaces

The concept of a neighbourhood space is a considerable generalisation of that of a metric space and, as we have seen, it allows meaningful, set-theoretic based, ideas of 'nearness' and 'convergence'. However, the collective experience of the mathematical community is that this concept needs to be supplemented with an additional requirement in order to yield a really useful tool.

The problem with a general neighbourhood space is the absence of any generic relation between the filters $\mathcal{N}(x)$ at different points $x$ in $X$. The crucial extra requirement that has emerged over the years is that any neighbourhood of a point $x$ should also be a neighbourhood of all points 'sufficiently near' to $x$ (and, of course, the notion of 'sufficiently near' must itself be defined in terms of the neighbourhoods of $x$). It is intriguing to ponder whether or not this is also relevant for a mathematical model that is to describe *physical* spacetime. For example, this is certainly what is assumed in classical general relativity, but it is by no means obvious that it should also hold sway in the quantum realm.

The precise mathematical definition of the new structure is as follows.

**Definition 1.10**

> A *topological space* is a neighbourhood space $(X, \mathcal{N})$ in which, for all $x \in X$ and for all $N \in \mathcal{N}(x)$, there exists $N_1 \in \mathcal{N}(x)$ such that, for all $y \in N_1$, $N \in \mathcal{N}(y)$. This is sketched in Figure 1.5.

The following theorem [Exercise!] is of considerable importance:

**Theorem 1.2** *A neighbourhood space $(X, \mathcal{N})$ is a topological space if and only if each filter $\mathcal{N}(x)$ has a filter base consisting of open sets.*

Thus, in a topological space, the neighbourhoods are essentially determined by the open sets alone; this is not true in a more general neighbourhood space.

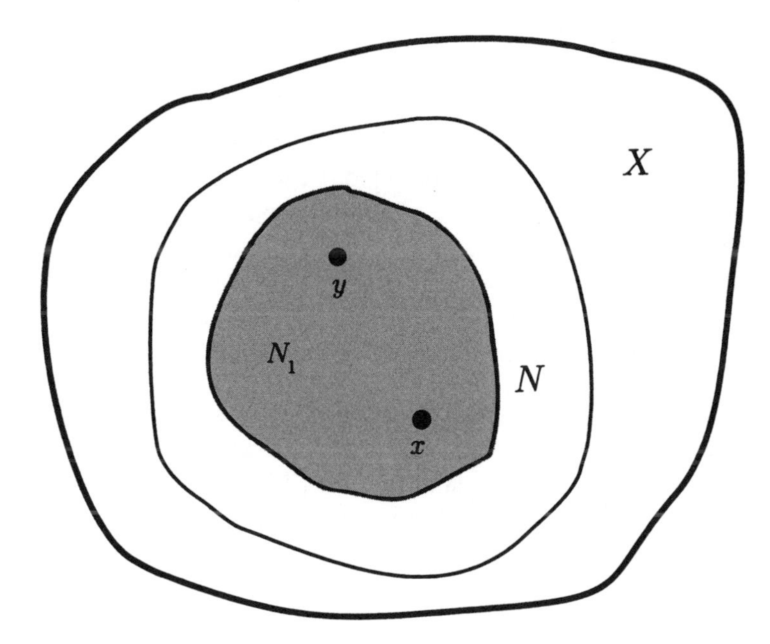


Figure 1.5: Neighbourhoods in a topological space.

**Examples**

1. Every metric or pseudo-metric space is a topological space since the balls $B_\epsilon(x) := \{ y \in X \mid d(x,y) < \epsilon \}$ are open [Exercise!] and form a basis for the neighbourhoods of $x$.

These spaces have many important properties. For example, every metric space is *first countable*, *i.e.,* there exists a countable basis for the neighbourhoods of each point (for example, the set of all balls whose radius is a rational number).

Note, however, that a metric space with a finite number of points is rather uninteresting: any such space automatically possesses the *discrete* topology [Exercise!] defined as the topology in which, for all $x \in X$, $\mathcal{N}(x)$ is the collection of all subsets of $X$ that contain the point $x$ (see below).

2. In the neighbourhood structure associated with pointwise convergence on the space of functions $\mathcal{F}([a,b],\mathbb{R})$, the sets $N_{t,\epsilon}(f)$ defined in Eq. (1.4.4) are open [Exercise!] and therefore so are their

finite intersections. But these form a filter base for $\mathcal{N}(f)$, and hence this function space is a topological space.

3. On a finite set $X$ with[24] $|X| = n$ there are $2^{n^2-n}$ different neighbourhood structures [Exercise!] but the number of topologies is less. For example, on $X = \{a, b, c\}$ there are 64 neighbourhood structures, but only 29 topologies. An example of a neighbourhood structure on $X = \{a, b, c\}$ that is not a topology is $\mathcal{N}(a) = \{\{a, b\}, X\}$, $\mathcal{N}(b) = \{\{b, c\}, X\}$ and $\mathcal{N}(c) = \{\{a, c\}, X\}$: the only open sets are $\emptyset$ and $X$, so they cannot form a basis for the neighbourhoods. □

We have seen that the collection of all open sets in a neighbourhood space satisfies the three conditions: (i) $\emptyset$ and $X$ are open; (ii) an arbitrary union of open sets is open; and (iii) any finite intersections of open sets is open. One of the central properties of a topological space is that the *converse* is true in the sense of the following theorem (whose proof is left as an exercise).

**Theorem 1.3** *Let $\tau$ be any family of subsets of a set $X$ that satisfies the three conditions:*

($\tau$1) *$\emptyset$ and $X$ belong to $\tau$;*

($\tau$2) *an arbitrary union of elements of $\tau$ belongs to $\tau$;*

($\tau$3) *any finite intersection of elements of $\tau$ belongs to $\tau$.*

*Then $\tau$ is the family of open sets of a topology on $X$ with a neighbourhood base $\mathcal{B}(x) := \{ O \in \tau \mid x \in O \}$ for all $x \in X$.*

### Comments

1. Let $\mathcal{N}(X)$ and $\tau(X)$ denote respectively the set of all neighbourhood structures on $X$ and the set of all topologies on $X$. Then $\tau(X) \subset \mathcal{N}(X)$ with an injection $i : \tau(X) \to \mathcal{N}(X)$ where, for a given topology $\tau$ on $X$, the neighbourhood structure $i(\tau)$ is obtained by defining the neighbourhood filter $N(x)$ of any $x \in X$ to be the collection of all subsets $N$ of $X$ that contain a $\tau$-open set that contains $x$;

[24]The number of elements in a finite set $X$ is denoted $|X|$.

*i.e.*, $N \in N(x)$ if and only if there exists $O \in \tau$ such that $x \in O \subset N$.

2. The open sets of an *arbitrary* neighbourhood structure $\mathcal{N}$ on $X$ obey the conditions in the theorem above and therefore generate a topological space associated with $\mathcal{N}$. This defines a map $k : \mathcal{N}(X) \to \tau(X)$, and it can be shown that, for any topology $\tau$, in the context of the diagram

$$\tau(X) \xrightarrow{i} \mathcal{N}(X) \xrightarrow{k} \tau(X), \tag{1.4.12}$$

we have $k \circ i(\tau) = \tau$.

3. The axioms ($\tau$1)–($\tau$3) constitute a complete, alternative, way of defining what is meant by a 'topology' on a set $X$. Indeed, many introductions to general topology start at this point by defining a topology on a space $X$ to be a collection $\tau$ of subsets of $X$ that satisfies the conditions ($\tau$1)–($\tau$3). For this reason, a topological space will be denoted by $(X, \tau)$ rather than $(X, \mathcal{N})$.

4. An equivalent way of defining a topological structure on a set $X$ is as a collection $\mathcal{C}$ of subsets of $X$ that

(i) include $\emptyset$ and $X$;

(ii) are algebraically closed under arbitrary intersections;

(iii) are algebraically closed under finite unions.

The complements of the elements in this family then form the collection of open sets for a unique topology on $X$ in which the original collection $\mathcal{C}$ is the family of closed sets. Thus a topology is also determined by its collection of closed sets, *i.e.*, sets that contain all their limit points.

5. As in the case of neighbourhood structures, it is convenient to introduce the notion of a 'base', or 'subbase', for a topology. Thus a collection $\mathcal{B}$ of subsets of $X$ is said to be a *base* for a topology $\tau$ if every $\tau$-open set can be written as a union of members of $\mathcal{B}$.

A collection of subsets is said to be a *subbase* if the set of all finite intersections of elements of the collection forms a base for the topology. It should be noted [Exercise!] that given *any* family $\mathcal{C}$ of subsets of a set $X$, there exists a unique 'smallest' topology on $X$ for which $\mathcal{C}$ is a subbase: namely that topology whose open sets are defined to be all arbitrary unions of the collection of all finite intersections of elements of $\mathcal{C}$.

6. In practice, topologies are almost always defined in terms of their open sets, or of a base or subbase for the open sets. For example, an interesting topology on any infinite set $X$ is the *cofinite* topology defined to be the one whose open sets are $\emptyset$ and the complements of all finite subsets of $X$.

7. Another important example is the 'product topology' on the Cartesian product $X_1 \times X_2$ of a pair of topological spaces $(X_1, \tau_1)$ and $(X_2, \tau_2)$. This topology is defined to be the union of all sets of the form $O_1 \times O_2$ where $O_1 \subset X_1$ is $\tau_1$-open and $O_2 \subset X_2$ is $\tau_2$-open. Thus a basis for this topology is precisely the collection of all sets of the form $O_1 \times O_2$.

The definition of the product topology admits a wide-ranging generalisation to the product of topological spaces $X_i$, $i \in I$, where $I$ is an *arbitrary* index set; here, the product $\times_{i \in I} X_i$ is defined to be the set of functions from $I$ to $\cup_{i \in I} X_i$ with the property that $f(i) \in X_i$ for all $i \in I$. Namely, the *product topology* on $\times_{i \in I} X_i$ is defined by specifying a subbase to be the collection

$$U(j, O) := \{f \in \times_{i \in I} X_i \mid f(j) \in O\} \tag{1.4.13}$$

where $j \in I$, and $O$ is open in the topology on $X_j$. Thus a basis for the product topology is the collection

$$\begin{aligned} &U(j_1, j_2, \ldots j_n; O_1, O_2, \ldots O_n) := \\ &\qquad \{f \in \times_{i \in I} X_i \mid f(j_1) \in O_1, f(j_2) \in O_2, \ldots f(j_n) \in O_n\} \end{aligned} \tag{1.4.14}$$

for all finite collections $\{j_1, j_2, \ldots j_n\}$ of indices, and where each subset $O_{j_i} \subset X_{j_i}$, $i = 1, 2, \ldots n$, is an open set in the topology on $X_{j_i}$.

8. Partially-ordered sets possess several natural topologies related to their ordering structure. One of the simplest is the collection of all upper sets, which is clearly a topology since it is algebraically closed

under arbitrary unions and intersections. The collection of all lower sets yields another example.

9. Interior points, exterior points, and boundary points have the following key properties for a subset $A$ of a topological space $X$:

- A point $x \in X$ is an *interior point* of $A$ if, and only if, there exists an open set $O$ such that $x \in O \subset A$. The interior of $A$—defined to be the set $\text{Int}(A)$ of all interior points in $A$—is the 'largest' open subset of $A$; more precisely, it is the union of all the open subsets of $A$ (that this union is indeed an open set, follows from the property than an arbitrary union of open sets is open).
- A point $x \in X$ is an *exterior point* of $A$ if, and only if, there exists an open set $O$ with $x \in O$ and $O \cap A = \emptyset$. This is equivalent to saying that $x$ is an interior point of the complement of $A$.
- A point $x \in X$ is a boundary point of $A$ if there exists an open set $O$ with $x \in O$, and $O \cap A \neq \emptyset$ and $O \cap A^c \neq \emptyset$.

An associated idea is of the *closure* $\bar{A}$ of a subset $A$ of a topological space $X$. This is defined to be the 'smallest' closed subset of $X$ that contains $A$; more precisely, it is the intersection of all the closed subsets of $X$ that contain $A$ (that this intersection is indeed a closed set follows from the property that an arbitrary intersection of closed sets is closed).

### 1.4.4 Some examples of topologies on a finite set

The idea of a topology can be illustrated with the aid of examples on a set $X$ that has only a finite number of elements.

**Problem 1.**

Show that the collection

$$\tau := \{\emptyset, X, \{a\}, \{c,d\}, \{a,c,d\}, \{b,c,d,e\}\} \tag{1.4.15}$$

defines a topology on the set $X := \{a,b,c,d,e\}$.

i) List the closed subsets of $X$. Hence show that there are subsets of $X$ that are neither open or closed; and also subsets of $X$ that are both open and closed.

ii) What are the closures of the sets $\{a\}$, $\{a, c\}$ and $\{b, d\}$?

iii) Let $A$ be the subset $\{b, c, d\}$ of $X$. Find the interior and exterior points of $A$, and hence find the boundary of $A$ in $X$.

**Answer 1.**

We have to show that the collection of subsets in Eq. (1.4.15) of the set $X := \{a, b, c, d, e\}$ satisfies the three axioms $(\tau_1)$–$(\tau_3)$:

**$(\tau_1)$:** $\emptyset$ and $X$ belong to $\tau$ by definition.

**$(\tau_2)$:** The calculation of the unions of the members of $\tau$ is as follows:

$$\begin{aligned}
&\{a\} \cup \{c, d\} = \{a, c, d\} \in \tau \\
&\{a\} \cup \{a, c, d\} = \{a, c, d\} \in \tau \\
&\{a\} \cup \{b, c, d, e\} = \{a, b, c, d, e\} = X \in \tau \\
&\{c, d\} \cup \{a, c, d\} = \{a, c, d\} \in \tau \\
&\{c, d\} \cup \{b, c, d, e\} = \{b, c, d, e\} \in \tau \\
&\{a, c, d\} \cup \{b, c, d, e\} = \{a, b, c, d, e\} = X \in \tau.
\end{aligned} \tag{1.4.16}$$

**$(\tau_3)$:** The calculation of the intersections of the members of $\tau$ is as follows:

$$\begin{aligned}
&\{a\} \cap \{c, d\} = \emptyset \in \tau \\
&\{a\} \cap \{a, c, d\} = \{a\} \in \tau \\
&\{a\} \cap \{b, c, d, e\} = \emptyset \in \tau \\
&\{c, d\} \cap \{a, c, d\} = \{c, d\} \in \tau \\
&\{c, d\} \cap \{b, c, d, e\} = \{c, d\} \in \tau \\
&\{a, c, d\} \cap \{b, c, d, e\} = \{c, d\} \in \tau.
\end{aligned} \tag{1.4.17}$$

Thus all the conditions for $\tau$ to be a topology are satisfied.

**i)** A closed set is defined to be any subset of $X$ that is the complement of an open set. The open sets are

$$\emptyset, X, \{a\}, \{c,d\}, \{a,c,d\}, \{b,c,d,e\} \tag{1.4.18}$$

and hence the closed sets are

$$X, \emptyset, \{b,c,d,e\}, \{a,b,e\}, \{b,e\}, \{a\}. \tag{1.4.19}$$

An example of a subset of $X$ that is neither open nor closed is $\{a,b\}$.

An example of a subset of $X$ that is both open and closed is $\{a\}$.

**ii)** The closure $\bar{A}$ of a subset $A \subset X$ is defined to be the intersection of all the closed subsets $F$ of $X$ such that $A \subset F \subset X$; *i.e.*, it is the 'smallest' closed subset of $X$ that contains $A$. Thus, by inspection,

$$\begin{aligned} \overline{\{a\}} &= \{a\} \\ \overline{\{a,c\}} &= \{a,b,c,d,e\} = X \\ \overline{\{b,d\}} &= \{b,c,d,e\}. \end{aligned} \tag{1.4.20}$$

**iii)** We have $A := \{b,c,d\}$. A point $x \in X$ is an interior point of $A$ if there exists an open set $O$ such that $x \in O \subset A$. However, the only non-empty, open subset of $A$ is $\{c,d\}$; hence the interior points of $A$ are $c$ and $d$.

A point $x \in X$ is an exterior point of $A$ if there exists an open set $O$ such that $x \in O \in A^c$. However, $A^c = \{a,e\}$; and the only open subset of this is $\{a\}$. Hence the only exterior point of $A$ is $a$.

A point $x \in X$ is a boundary point of $A$ if it is neither an interior point nor an exterior point. Hence the boundary points are $b$ and $e$; and hence $\text{bd}(A) = \{b,e\}$.

**Problem 2.**

Let $X$ denote the finite set $\{a,b,c,d,e\}$. Which of the following classes of subsets constitutes a topology on $X$?

i) $\{\emptyset, X, \{a\}, \{a,b\}, \{a,c\}\}$

ii) $\{\emptyset, X, \{a,b,c\}, \{a,b,d\}, \{a,b,c,d\}\}$

iii) $\{\emptyset, X, \{a\}, \{a,b\}, \{a,c,d\}, \{a,b,c,d\}\}$.

**Answer 2.**

i) The set $\tau_1 := \{\emptyset, X, \{a\}, \{a,b\}, \{a,c\}\}$ is not a topology on $X$ since $\{a,b\}$ and $\{a,c\}$ belong to $\tau_1$, but the union $\{a,b\} \cup \{a,c\} = \{a,b,c\}$ is not a member of the collection $\tau_1$.

ii) The set $\tau_2 := \{\emptyset, X, \{a,b,c\}, \{a,b,d\}, \{a,b,c,d\}\}$ is not a topology on $X$ since $\{a,b,c\}$ and $\{a,b,d\}$ belong to $\tau_2$, but the intersection $\{a,b,c\} \cap \{a,b,d\} = \{a,b\}$ is not a member of the collection $\tau_2$.

iii) The set $\tau_3 := \{\emptyset, X, \{a\}, \{a,b\}, \{a,c,d\}, \{a,b,c,d\}\}$ is a topology on $X$ since, by inspection, this collection of subsets of $X$ is algebraically closed under the operations of intersection and union (and it contains $\emptyset$ and $X$).

### 1.4.5 A topology as a lattice

It is interesting to note that, for a given topology $\tau$ on $X$, the set of all $\tau$-open sets is a lattice. The key definitions are as follows.

**Definition 1.11**

1. If $O_1, O_2$ are $\tau$-open sets, the partial-ordering operation is defined by $O_1 \preceq O_2$ if and only if $O_1 \subset O_2$.

2. The *meet* of two $\tau$-open sets $O_1, O_2$ is defined as

$$O_1 \wedge O_2 := O_1 \cap O_2. \tag{1.4.21}$$

3. The *join* of two $\tau$-open sets $O_1, O_2$ is defined as

$$O_1 \vee O_2 := O_1 \cup O_2. \tag{1.4.22}$$

4. The *unit* element 1 and *null* element 0 are defined as

$$1 := X \text{ and } 0 := \emptyset \tag{1.4.23}$$

respectively.

5. The *pseudo-complement* of the $\tau$-open set $O$ is defined as

$$O' := \operatorname{Int}(X - O). \tag{1.4.24}$$

**Comments**

1. The partial-ordering, meet and join operations are the same as those on the lattice $P(X)$ of *all* (*i.e.*, not just open) subsets of $X$. The lattice $\tau$ inherits the distributive nature of the meet and join operations of $P(X)$.

2. The lattice $\tau$ is not complete since it is not necessarily closed under the operation of taking arbitrary intersections (all that is guaranteed by the axioms of topology is that the lattice of open sets is closed under the operation of taking *finite* intersections of its elements). If desired, this can be remedied by defining the meet of an arbitrary family $O_i$, $i \in I$, of open sets to be the open set

$$\wedge_{i\in I} O_i := \operatorname{Int} \bigcap_{i\in I} O_i. \tag{1.4.25}$$

3. The pseudo-complement is generally not an orthocomplement (in the sense that $(A')' = A$) since one typically has

$$O \prec (O')' \tag{1.4.26}$$

with $O \neq (O')'$.

For example, consider the topology $\{\emptyset, \{a\}, X\}$ on the set $X := \{a, b, c\}$. Then $\{a\}' = \operatorname{Int}(X - \{a\}) = \operatorname{Int}(\{b, c\}) = \emptyset$. Hence $\{a\}'' = \emptyset' = \operatorname{Int}(X - \emptyset) = \operatorname{Int}(X) = X$. Thus $\{a\}''$ is not equal to $\{a\}$.

Similarly, we may have the strict inequality

$$O \vee O' \prec 1. \tag{1.4.27}$$

For example, in the topology $\{\emptyset, \{a\}, X\}$ on $X := \{a, b, c\}$, we have $\{a\} \vee \{a\}' = \{a\} \cup \emptyset = \{a\}$ which is a proper subset of $X = \{a, b, c\}$.

As another example, consider $O \vee O'$ when $O$ is the unit disk $O := \{z \in \mathbb{C} \mid |z| < 1\}$ in the complex plane $\mathbb{C}$ equipped with its standard metric topology. Clearly $O' = \{z \in \mathbb{C} \mid |z| > 1\}$, and hence $O \cup O'$ is equal to $\mathbb{C}$ minus the unit circle $\{z \in \mathbb{C} \mid |z| = 1\}$.

4. In the lattice $P(X)$ of *all* subsets of $X$ we *do* have both the relations $A = A'$ and $A \vee A' = X$. This is one of the main reasons why lattices of this type can serve as mathematical models of classical logic.

On the other hand, in what is called *intuitionistic* logic, one typically has $\alpha \prec \alpha''$ and $\alpha \vee \alpha' \prec 1$. Thus topological spaces give natural models for this type of logic. This connection between intuitionistic logic and topology finds its deepest expression in the subject of *topos* theory. The analogue of the idea of a Boolean algebra in this situation is what is known as a *Heyting* algebra (MacLane & Moerdijk 1992).

### 1.4.6 The lattice of topologies $\tau(X)$ on a set $X$

One of the very interesting properties of the set $\tau(X)$ of all topologies on a set $X$ is that it can be equipped with a natural lattice structure (this must not be confused with the statement in the previous section that the collection of all open sets for a *given* topology on $X$ is lattice). The first step is to make $\tau(X)$ into a poset:

**Definition 1.12**

1. The set $\tau(X)$ of all topologies on a set $X$ can be partially ordered, with $\tau_1 \preceq \tau_2$ defined to mean that every $\tau_1$-open set is $\tau_2$-open (that is, $\tau_2$ has 'more' open sets than $\tau_1$).

2. In these circumstances, the topology $\tau_1$ is said to be *weaker*, or *coarser*, than $\tau_2$; and $\tau_2$ is *stronger*, or *finer*, than $\tau_1$. This notation is compatible with the earlier use of 'stronger' and 'weaker' in relation to metrics.

3. The strongest topology is $P(X)$ (*i.e.*, every subset of $X$ is open) and is called the *discrete* topology.

   The weakest topology (the *indiscrete* topology) is just $\{\emptyset, X\}$.

The set of all topologies $\tau(X)$ on $X$ can be given a lattice structure with respect to which this ordering is the lattice ordering. The lattice operations on $\tau(X)$ are defined as follows:

**Definition 1.13**

1. The *meet* operation is

$$\tau_1 \wedge \tau_2 := \tau_1 \cap \tau_2 = \{A \subset X \mid A \text{ is open in both } \tau_1 \text{ and } \tau_2\}. \tag{1.4.28}$$

2. The *join* operation is

$$\tau_1 \vee \tau_2 := \text{ coarsest topology containing } \{A_1 \cap A_2 \mid A_1 \in \tau_1, A_2 \in \tau_2\}. \tag{1.4.29}$$

3. The *null* and *unit* elements are respectively the weakest topology $\{\emptyset, X\}$ and the strongest topology $P(X)$.

**Comments**

1. It is instructive to study a few simple examples where $X$ is a finite set. The number of topologies that can be placed on a given set $X$ has been calculated for the cases $|X| = 1 - 7$ and is equal to 1,4,29,355,6942,209527 and 9535241 respectively.

In general, if $|X|$ is a finite integer $n$, it is known that $2^n \leq |\tau(X)| \leq 2^{n(n-1)}$. When $X$ is infinite it can be shown (Fröhlich 1964) that the cardinality of $\tau(X)$ is two orders of infinity higher than that of $X$ (that is, $|\tau(X)| = 2^{2^{|X|}} \equiv |P(P(X))|$).

2. The simplest case is when $X$ has one element, $\{a\}$ say, for which there is just the single topology $\{\emptyset, \{a\}\}$.

If $X = \{a, b\}$ there are four topologies arranged in the following lattice:

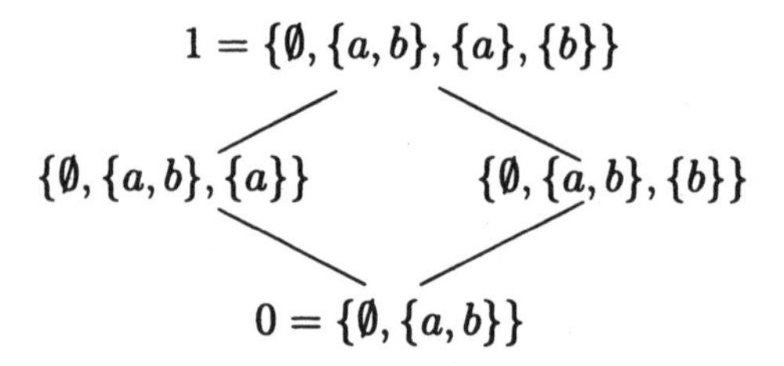


3. The first really interesting example is when $X$ is a set $\{a, b, c\}$ of cardinality 3. The lattice diagram for this case is shown in Figure 1.6 using a notation that has been chosen for maximum typographical simplicity. For example, $ab(ab)(ac)$ means the topology whose open sets other than $\emptyset$ and $X$ are the subsets $\{a\}$, $\{b\}$, $\{a, b\}$ and $\{a, c\}$.

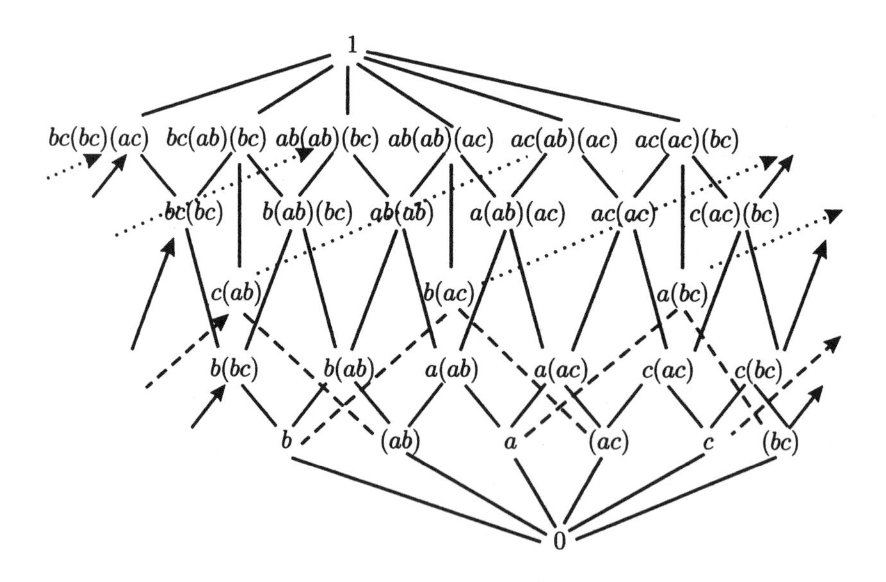


Figure 1.6: The lattice of topologies for $X = \{a, b, c\}$.

4. The lattice of all topologies on $X$ possesses many interesting properties (Larson & Andima 1975). For example, it is complete and *atomic*. More precisely, for each $A \subset X$, $\tau_A = \{\emptyset, X, A\}$ is an atom

(that is, $\tau_A$ covers the trivial topology 0), and every topology $\tau$ is determined by these atoms in the sense that

$$\tau = \bigvee \{\tau_A \mid \tau_A \preceq \tau\}. \tag{1.4.30}$$

In the example above where $X = \{a, b, c\}$, the atoms are the six topologies $b$,$(ab)$,$a$,$(ac)$,$c$ and $(bc)$.

5. The lattice $\tau(X)$ is also *anti-atomic*. That is, there exist topologies $\tau_A$ with the properties that (i) the maximal topology 1 covers $\tau_A$; and (ii) every topology is uniquely determined by the anti-atoms that lie above it.

In the example where $X = \{a, b, c\}$, the anti-atoms are the topologies $bc(bc)(ac)$, $bc(ab)(bc)$, $ab(ab)(bc)$, $ab(ab)(ac)$, $ac(ab)(ac)$ and $ac(ac)(bc)$. In general, the anti-atoms are topologies of the form

$$\tau_{(x,\mathcal{U})} := \{A \subset X \mid x \notin A \text{ or } A \in \mathcal{U}\} \tag{1.4.31}$$

where $\mathcal{U}$ is any ultra-filter (a maximal element with respect to the natural partial ordering of filters) not equal to the principal ultra-filter of all subsets of $X$ containing the point $x \in X$.

### 1.4.7 Some properties of convergence in a general topological space

It is important to note that if $x_n$ is a sequence in $A \subset X$ that converges to $x$, then $x$ is a limit point of $A$. However, unlike the situation for metric spaces, the converse may not be true: a subset $A$ of a general topological space may have a limit point to which no sequence of elements in $A$ converges. But what *is* true is that there will always be a *filter* on $A$ that converges to the limit point.

Thus a general topological structure is determined by its collection of convergent filters. It can be shown that a necessary and sufficient condition for a topology to be determined by the set of convergent sequences alone is that it be first countable (Bourbaki 1966, Kelly 1970).

Generalised convergence can also be discussed using what are called 'nets':

**Definition 1.14**

1. A *directed set* is a partially-ordered set $D$ with the property that if $\alpha, \beta \in D$ then there exists $\gamma \in D$ such that $\alpha \preceq \gamma$ and $\beta \preceq \gamma$.

2. A *net* on a set $X$ is any function $f : D \rightarrow X$ for some directed set $D$.

3. A net on $X$ *converges* to a point $x \in X$ with respect to a neighbourhood structure $\mathcal{N}$ on $X$ if

$$\text{“for all } N \in \mathcal{N}(x), \text{ there exists } \alpha \in D \text{ such that } \gamma \succeq \alpha \text{ implies } f(\gamma) \in N\text{”}. \quad (1.4.32)$$

The set of all positive integers is a special example of a directed set, and hence Eq. (1.4.32) is a far-reaching generalisation of the definition Eq. (1.4.11) of a convergent sequence. It can be shown that there is a one-to-one correspondence between filters and nets, and therefore nothing is lost by using the latter. Many authors prefer this (for example, Kelly (1970)) because of the intuitive similarity between nets and sequences. However, I have elected to concentrate on filters because their definition as 'dual ideals' in the lattice $\mathrm{P}(X)$ is in line with my desire to emphasise some of the algebraic aspects of general topology.

### 1.4.8 The idea of a compact space

A most important concept in topology—and one that fundamentally involves generalised convergence—is that of a 'compact' space, which means a space that is, in some sense, of 'finite size'. The classic examples of compact spaces are spheres, tori, or any other subspaces of Euclidean space $\mathbb{R}^n$ that are closed and bounded[25].

[25] A subspace $A$ of a metric space is *bounded* if $\sup_{x,y\in A} d(x,y) < \infty$.

One characteristic feature of such a set is that any infinite subset of points must necessarily cluster together in some way. More precisely, it can be shown that every sequence $(x_1, x_2, \ldots)$ in a closed and bounded subset of $\mathbb{R}^n$ necessarily has at least one *accumulation* point, defined as any point $x$ such that any neighbourhood of $x$ is visited infinitely many times by the sequence:

$$\text{“for all } N \in \mathcal{N}(x), \text{ for all } n, \text{ there exists } n' > n \text{ such that } x_{n'} \in N\text{”} \tag{1.4.33}$$

or, in terms of the tails $T_n$ of the sequence,

$$\text{“for all } N \in \mathcal{N}(x), \text{ for all } T_n, \ \mathcal{N} \cap T_n \neq \emptyset\text{”}. \tag{1.4.34}$$

One might try and reverse this result and *define* a general compact space to be any topological space in which Eq. (1.4.34) is true. However, it turns out that this is too broad, and the most useful definition is to strengthen Eq. (1.4.34) by including *all* filter bases, not just the tails of sequences:

**Definition 1.15**

A topological space $X$ is *compact* if every filter base $\mathcal{B}$ on $X$ has an accumulation point. That is, there exists $x \in X$ such that

$$\text{“for all } N \in \mathcal{N}(x), \text{ for all } A \in \mathcal{B}, \ N \cap A \neq \emptyset\text{”}. \tag{1.4.35}$$

**Comments**

1. An alternative, well-known definition involves properties of ‘coverings’ of $X$ by families of open sets; the reader is referred to one of the cited standard texts for further discussion of this idea.

2. A famous result in topology is the *Heine-Borel* theorem which has been alluded to already. This asserts that every closed and bounded subset of the Euclidean space $\mathbb{R}^n$ is compact—a result that is extremely useful in practice.

3. Another famous theorem is the *Tychonoff product theorem* which asserts that the product topology on the product of an arbitrary family of compact spaces is itself compact.

### 1.4.9 Maps between topological spaces

A crucial concept in most branches of mathematics is of a structure-preserving map between two sets equipped with the same type of mathematical structure; in our case this is essentially the lattice of open sets associated with a topology. The first relevant question in the present context, therefore, is whether a map $f : X \to Y$ between a pair of sets $X$ and $Y$ induces any maps between $P(X)$ and $P(Y)$ that respect the lattice structure. From a purely set-theoretic perspective, there are two natural maps—one from $P(X)$ to $P(Y)$, and one from $P(Y)$ to $P(X)$. These are defined as follows.

**Definition 1.16**

1. The *induced* map from $P(X)$ to $P(Y)$ is defined on a subset $A \subset X$ by
$$f(A) := \{\, f(x) \in Y \mid x \in A \,\} \tag{1.4.36}$$
and has the properties
$$\text{(a)} \quad f(A \cup B) = f(A) \cup f(B); \tag{1.4.37}$$
$$\text{(b)} \quad f(A \cap B) \subset f(A) \cap f(B). \tag{1.4.38}$$
Note that the equality may not hold in Eq. (1.4.38), and hence the induced map from $P(X)$ to $P(Y)$ does *not* preserve the lattice structure.

2. The second map is from $P(Y)$ to $P(X)$, and is defined on $A \subset Y$ as the *inverse set map*
$$f^{-1}(A) := \{\, x \in X \mid f(x) \in A \,\} \tag{1.4.39}$$
which, it should be noted, is well-defined even if $f : X \to Y$ is not one-to-one. This map satisfies
$$\text{(a)} \quad f^{-1}(A \cup B) = f^{-1}(A) \cup f^{-1}(B) \tag{1.4.40}$$
$$\text{(b)} \quad f^{-1}(A \cap B) = f^{-1}(A) \cap f^{-1}(B) \tag{1.4.41}$$
with generalisations to arbitrary families. Hence it *does* preserve the lattice operations.

The results in Eqs. (1.4.40–1.4.41) motivate (and render consistent) the following definition of a 'continuous' map.

**Definition 1.17**

> A map $f : (X, \tau) \to (Y, \tau')$ between topological spaces is *continuous* if, for all $O \in \tau'$, $f^{-1}(O) \in \tau$.

**Comments**

1. It follows from Eqs. (1.4.40–1.4.41) that a continuous map between two topological spaces induces a homomorphism from the lattice $\tau'$ into the lattice $\tau$. Thus a continuous function is a structure-preserving map when the subject of topology is viewed from the perspective of the lattice of open sets. The significance of this will be touched on later in the context of the ideas of frames and locales.

2. A more intuitive idea of the continuity of a function $f$ is that a 'small variation' in $x$ produces only a small variation in the value $f(x)$. In the absence of a metric, the concept of 'small' must be defined in terms of the neighbourhoods of the points $x$ and $f(x)$: in fact, it can be shown that a function $f : (X, \tau) \to (Y, \tau')$ is continuous if and only if

$$\text{“for all } x \in X \text{ and } M \in \mathcal{N}(f(x)),\ \text{there exists } N \in \mathcal{N}(x) \text{ such that } f(N) \subset M\text{”} \tag{1.4.42}$$

or, equivalently, $\mathcal{N}(f(x)) \vdash f(\mathcal{N}(x))$. Note that when $X = Y = \mathbb{R}$ this reduces to the familiar definition:

$$\text{“for all } x \in \mathbb{R},\ \epsilon > 0,\ \text{there exists } \delta > 0 \text{ such that } |x - y| < \delta \text{ implies } |f(x) - f(y)| < \epsilon\text{”}. \tag{1.4.43}$$

3. The 'small' variation is often phrased in terms of sequences. Thus if a function $f : X \to Y$ is continuous and if $x_n \to x$ then $f(x_n) \to f(x)$ [Exercise!]. If $X$ is a metric space, the converse also holds. That is, if $f : X \to Y$ is such that, for all points $x \in X$ and for any convergent sequence $x_n \to x$, $f(x_n)$ converges to $f(x)$, then $f$ is continuous.

For more general spaces this is false, but what *is* true is that a function $f : (X, \tau) \to (Y, \tau')$ is continuous if and only if for all $x \in X$ and for any filter base $\mathcal{B}$ on $X$ that converges to $x$ it is true that $f(\mathcal{B})$ converges to $f(x)$. (Exercise: show that if $\mathcal{B}$ is a filter base on $X$ and $f$ is any map from $X$ to $Y$, then $f(\mathcal{B}) := \{ f(A) \mid A \in \mathcal{B} \}$ is a filter base on $Y$.)

4. One of the important practical problems in the theory of topology is to find ways of constructing 'natural' topologies on a given set. Two of the most useful techniques involve placing a topology on a set $X$ with the aid of maps to or from $X$ and some other topological space $Y$. For example, if $\tau$ is a topology on $Y$, and $f$ is a map from $X$ to $Y$, the *induced* topology on $X$ is defined to be

$$f^{-1}(\tau) := \{f^{-1}(O) \mid O \in \tau\}. \tag{1.4.44}$$

The key property of the induced topology is that it is the coarsest topology on $X$ such that $f$ is continuous. It is because of the existence of this special property that we refer to this topology as 'natural'.

A special case is when $X$ is a subset of $Y$ with an injection $i : X \to Y$. The induced topology on $X$ is then called the *subspace topology* and consists of all sets of the form $X \cap O$ where $O$ is open in the topology $\tau$ on Y. Note that the results Eqs. (1.4.40–1.4.41) are crucial for the success of this construction.

5. Another important example arises when $(Y, \tau)$ is a topological space and there is a surjective map $p : Y \to X$. The *identification topology* on $X$ is defined as

$$p(\tau) := \{A \subset X \mid p^{-1}(A) \in \tau\}. \tag{1.4.45}$$

The key property of this topology is that it is the finest one on $X$ such that $p$ is continuous.

A common example of such a situation is when some equivalence relation $R$ is defined on a set $Y$, and $X$ is the space $Y/R$ of equivalence classes with $p$ being the canonical map of an element of $Y$ onto its equivalence class.

In fact, this example is universal since if $p$ is *any* surjective map from a space $Y$ onto a set $X$, an equivalence relation can be defined on

$Y$ by saying that two points $y_1$ and $y_2$ are equivalent if $p(y_1) = p(y_2)$. It is then easy to see that a bijection can be established between $X$ and $Y/R$ by mapping the point $p(y) \in X$ to the equivalence class of $y$ in $Y$. Note that the points in $Y$ that are mapped into the same point in $Y/R$ are precisely those that are equivalent to each other. Hence one says that $Y/R$ is obtained by 'identifying equivalent points'; this explains the origin of the name 'identification topology'.

### 1.4.10 The idea of a homeomorphism

A crucial question is when two topological spaces can be regarded as being equivalent. More precisely, we are looking for the appropriate meaning of an 'isomorphism' in the topological case.

Generally speaking, an isomorphism between two structures of the same type involves a bijective map between the underlying sets with the property that both it and its inverse are structure preserving. In the context of topology—working from the idea of a continuous map as a structure-preserving map—this suggests the following definition.

**Definition 1.18**

A map $f : (X, \tau) \to (Y, \tau')$ is a *homeomorphism* (an isomorphism in the context of general topology) if

(a) $f$ is a bijection;

(b) $f$ and $f^{-1}$ are continuous.

We shall write this as $(X, \tau) \simeq (Y, \tau')$. Note that the symbol $f^{-1}$ refers here to the map from $Y$ to $X$ that is the actual inverse of the map $f$ from $X$ to $Y$. It should not be confused with the inverse *set* map defined in Eq. (1.4.39). When $f$ is invertible the two maps are related by $\{f^{-1}(y)\} = f^{-1}(\{y\})$ for all $y \in Y$.

**Comments**

1. A bijection $f$ is a homeomorphism if and only if (i) for all $O \in \tau$, $f(O)$ is $\tau'$-open; and (ii) for all $O \in \tau'$, $f^{-1}(O)$ is $\tau$-open. Thus $f$ induces a bijective map between the collections of open sets

for the two topologies and with the property that it preserves the algebraic operations of forming unions and intersections. Thus the induced map is an isomorphism of the lattice structures associated with the two topologies.

2. The set Perm$(X)$ of all bijections ('permutations') of $X$ onto itself is a group. If $\tau$ is a topology on $X$ and $\phi \in \text{Perm}(X)$, $\phi(\tau)$ is defined to be the topology whose open sets are $\{\, \phi(O) \mid O \in \tau \}$. By construction, $\tau \simeq \phi(\tau)$ (*i.e.*, they are homeomorphic) and, conversely, if $\tau_1$ and $\tau_2$ are a pair of topologies on the same set $X$ that are homeomorphic then there exists $\phi \in \text{Perm}(X)$ such that $\tau_2 = \phi(\tau_1)$. Thus the set $\tau(X)$ of all topologies on $X$ decomposes under the action of Perm$(X)$ as a disjoint union of orbits that are the homeomorphism classes of topology. □

### 1.4.11 Separation axioms

An important question in any topological space $X$ is the extent to which points can be distinguished from each other by listing the collection of open sets to which each belongs.

From the viewpoint of conventional physics this is related to the idea that if $X$ represents physical space, then any real 'object' exists inside an open set. More precisely, it cannot exist as a subset of a closed subset unless this has a non-trivial interior.[26] In the context of quantum field theory, this remark is related to the analysis by Bohr and Rosenfeld[27] of the need to smear quantum fields with test functions that are non-vanishing on an open set. It thus seems plausible to argue that it is physically meaningless to distinguish between two points in $X$ if the collections of open sets to which they belong are identical. The relevant mathematical definitions for handling this type of consideration are as follows.

---

[26]One could say that all open sets are 'fat' whereas closed sets come in both thin and fat varieties. For example, a segment of a line in the plane is thin whereas a closed disc is fat.

[27]The relevant papers are conveniently reprinted and translated in Wheeler & Zurek (1983).

**Definition 1.19**

1. A topological space $X$ is $T_0$ if, given any pair of points $x, y \in X$, at least one of them is contained in an open set that excludes the other. This is equivalent to saying that, for all $x, y \in X$, $\mathcal{N}(x) \neq \mathcal{N}(y)$.

2. The space is $T_1$ if, given any pair of points $x, y$ each one is contained in an open set that excludes the other.

3. The space is $T_2$, or *Hausdorff*, if for any pair of points $x, y \in X$ there exist open sets $O_1$ and $O_2$ such that $x \in O_1$, $y \in O_2$ and $O_1 \cap O_2 = \emptyset$.

**Comments**

1. As remarked earlier, the closure $\bar{A}$ of any subset $A$ of a topological space $X$ is defined to be the smallest closed set containing $A$ (it can be constructed as the intersection of all closed sets containing $A$). It is easy to see that $\mathcal{N}(x) = \mathcal{N}(y)$ if and only if $\overline{\{x\}} = \overline{\{y\}}$.

2. If $X$ is $T_0$, a partial ordering on $X$ can be defined by

$$x \preceq y \text{ if } \overline{\{x\}} \subset \overline{\{y\}}. \tag{1.4.46}$$

3. From the remarks made above, it could arguably be asserted that any topological space that represents spacetime must be at least $T_0$—at least, if all its points are to have 'physical meaning' in the sense of being distinguishable by objects located in open sets. It is important, therefore, to note that to *any* topological space $X$ there is an associated $T_0$ space. This is constructed by defining the equivalence relation $R$ on $X$

$$xRy \text{ if } \mathcal{N}(x) = \mathcal{N}(y) \tag{1.4.47}$$

and then equipping $X/R$ with the identification topology of Eq. (1.4.45). The resulting space is $T_0$, and can be regarded as the space that is obtained from the original space $X$ once points that cannot be 'physically separated' have been identified.

It is interesting to note that if this procedure is applied to a space with a pseudo-metric $\rho$, the resulting space is actually $T_2$, and with a metric topology that is induced by the distance function $d([x],[y]) := \rho(x,y)$ where $[x]$ denotes the equivalence class of the point $x$.

4. A space is $T_1$ if and only if, for every point $x \in X$, the subset $\{x\}$ of $X$ is closed.

Note that there exist topological spaces that are $T_0$ but not $T_1$. An example is the set $X = \{a,b\}$ with the topology $\{\emptyset, X, \{a\}, \{a,b\}\}$. Indeed, it is easy to see that the only topology on a finite set that is $T_1$ is the discrete topology $P(X)$.

5. An example of a topology that is $T_1$ but not $T_2$ is the cofinite topology on any infinite set $X$.

Note also that (i) every metric space is Hausdorff; but (ii) the only Hausdorff topology on a finite set is the discrete topology [Exercise!].

6. A subspace of a $T_i$ topology, $i = 0,1,2$ is also a $T_i$ topology.

7. There exist more refined notions of separation that involve, for example, the extension of the Hausdorff axiom to include the ability to distinguish between *arbitrary* closed sets (not merely single points) with the aid of non-intersecting open sets that contain them. Considerations of this type lead to the ideas of a $T_3$ topology or a $T_4$ topology.

8. The question of the *uniqueness* of the limits of sequences (or, more generally, filters) in a topological space has a precise answer in terms of the separation properties of the space. Specifically, it can be shown that a necessary and sufficient condition for a topological space $X$ to be Hausdorff is that every filter on $X$ converges to at most one point in $X$.

### 1.4.12 Frames and locales

I wish now to discuss briefly certain algebraic structures that are associated with any topological space and from which the topology can be largely reconstructed. One well-known example is that the topology of a compact Hausdorff space $X$ is uniquely specified by the ring structure of its set $C(X)$ of real-valued continuous functions.

Another example is based on the observation in Section 1.4.5 that the collection of subsets of $X$ that constitutes a topology forms a *sublattice* of $\mathrm{P}(X)$. This raises the interesting question of the extent to which this lattice structure determines the topology. In particular, given the lattice, can the topological space be reconstructed (up to homeomorphisms)?

From the discussion in Section 1.4.11 it seems unlikely that the topology can be reproduced completely if $X$ contains points that cannot be separated by specifying the open sets to which they belong: *i.e.*, if the topology is not $T_0$. For example, the topologies $\tau_1 := \{\emptyset, X, \{a\}\}$ and $\tau_2 := \{\emptyset, X, \{b, c\}\}$ on the set $X = \{a, b, c\}$ have isomorphic lattices of open sets but $\tau_1$ is not homeomorphic to $\tau_2$.

The basic step is to try to reconstruct the points of $X$ from the lattice associated with a topology $\tau$ on $X$. Since the only question that can be asked of a point is whether or not it belongs to any particular open set, it is natural to consider the collection of mappings $h_x : \tau \to \{0, 1\}$, $x \in X$, defined on open sets $O$ by

$$h_x(O) := \begin{cases} 1, & \text{if } x \in O; \\ 0, & \text{otherwise.} \end{cases} \tag{1.4.48}$$

It can be seen at once that each $h_x$ is a homomorphism from the lattice $\tau$ onto the lattice $\{0, 1\}$ of two points. This inspires an attempt to define a 'generalised point' associated with the lattice to be *any* homomorphism from the lattice into $\{0, 1\}$. A topology can be constructed on the set $\mathrm{pt}(\tau)$ of all such homomorphisms by defining the open sets to be all subsets of the form $\{h \in \mathrm{pt}(\tau) \mid h(O) = 1\}$ where $O$ is any $\tau$-open subset of X. The natural map $h : X \to \mathrm{pt}(\tau)$, defined by $x \mapsto h_x$, is clearly continuous with respect to this topology.

A number of important statements can be made concerning this construction:[28]

[28] A comprehensive discussion of the subject is Johnstone (1986); see also Vickers (1989).

1. The topology on $\mathrm{pt}(\tau)$ is $T_0$.

2. Two points $x, y \in X$ determine the same homomorphism if, and only if, $\mathcal{N}(x) = \mathcal{N}(y)$. Thus the map $h : X \to \mathrm{pt}(\tau)$ is one-to-one if and only if the topology $\tau$ on $X$ is $T_0$. If a relation $R$ is defined on $X$ as in Eq. (1.4.47), it is clear that the map $[x] \mapsto h_x$ is a continuous injection of $X/R$ into $\mathrm{pt}(\tau)$.

3. The map $x \mapsto h_x$ may not be surjective: *i.e.*, there may exist homomorphisms that are *not* of the form $h_x$ for any $x \in X$. Spaces for which this map is both one-to-one and onto (it is then necessarily a homeomorphism) are called *sober*—they are the spaces whose topology is completely captured by the lattice structure of their open sets. For example, all Hausdorff spaces are sober; the cofinite topology on an infinite set $X$ is not.

At this point it is important to observe that there is no reason why the constructions above cannot be applied to lattices that are not *a priori* lattices of open sets in any topology! To see what type of lattice is appropriate for such a treatment note that (i) the lattice of open sets associated with a topology is algebraically closed under arbitrary unions (the join operation); (ii) if the meet of an arbitrary family of open sets is defined to be the interior of their intersection, the lattice of open sets becomes a complete sublattice of $P(X)$; and (iii) the lattice of open sets obeys the infinite distributive law

$$A \wedge \bigvee S = \bigvee \{ A \wedge B \mid B \in S \} \tag{1.4.49}$$

where $S$ is any collection of open sets.

It is this collection of properties that is axiomatised to construct a purely algebraic definition of a topology-like structure. More precisely, a *frame* or *locale*[29] is defined to be any complete lattice that satisfies the infinite distributive law. Many of the ideas in topology generalise to this situation, and this has given rise to the interesting subject of 'pointless' topology (Johnstone, *loc cit*). In particular, the

[29]There is a technical difference between these two concepts that comes into play only when structure-preserving maps are being considered. See the literature for more details.

'points' associated with any frame $\ell$ are defined to be the homomorphisms from $\ell$ into $\{0, 1\}$; the set of all of them is then given the topology in which the open sets are all subsets of $\mathrm{pt}(\ell)$ of the form $\{h \in \mathrm{pt}(\ell) \mid h(a) = 1\}$ for some $a \in \ell$.

If locales/frames are to replace point-set topology there has to be some analogue of the important idea of a continuous map $f : (X, \tau_1) \to (Y, \tau_2)$ from one topological space to another. The key step here is the result mentioned in Section 1.4.9 that such a map induces a homomorphism from the lattice $\tau_2$ into the lattice $\tau_1$. Thus homomorphisms between locales replace the idea of continuous maps. In this context it is relevant to note that:

1. if $f$ is a one-to-one map, the associated lattice map is surjective; the converse holds if $\tau_1$ is a $T_0$ topology;

2. if $f$ is surjective, the associated lattice map is injective.

It is clear from these results how one might set about constructing the appropriate generalisations to frames/locales of the ideas of subspace and quotient space.

This algebraic generalisation of topology is rather fascinating and it is attractive to speculate that structures of this type might one day form an important ingredient in a proper understanding of the quantum theory of space and time.

# Chapter 2

# Differentiable Manifolds

## 2.1 Preliminary Remarks

The minimal mathematical representation of spacetime in theoretical physics would presumably be with some set $\mathcal{M}$ whose elements represent 'spacetime points'. But, by itself, a set has no structure other than what it contains, and the question arises naturally, therefore, as to what other mathematical structure should be imposed on $\mathcal{M}$. In practice, the idea has arisen that, at the very least, $\mathcal{M}$ should also be a topological space whose open sets represent certain privileged regions in the spacetime.

However, the topological spaces that have actually been used historically to represent spacetime (or space and time separately) are of a very special type. Namely, they have the property that it is possible to uniquely label any particular spacetime point by specifying the values of a finite set of real numbers, the number of which is identified as the 'dimension' of the space. Thus, in Newtonian physics, three-dimensional physical space is represented mathematically by the Euclidean space $\mathbb{R}^3$, and one-dimensional time is represented by $\mathbb{R}$; in special relativity, the combined notion of 'spacetime' is represented by the Euclidean space $\mathbb{R}^4$.

The use of such familiar mathematical models has many important implications, not the least of which is that *differentiation* can be defined, thus opening up the very fruitful idea that the dynamical evolution of a physical system can be modelled by differential equations defined on the spacetime. Also, of course, there is an explicit underlying topology on such spaces: namely, the metric-space topology induced by the usual metric function.

It is a matter of some philosophical and physical doubt whether such a use of real numbers is totally meaningful: the construction of the real numbers from the rationals is quite an abstract operation, and many physicists would be happy with the idea that employing real numbers in this way is an idealisation that might at some future time be replaced with a different picture of the world at very small distances (for example, in relation to the Planck length $L_P := (G\hbar/c^3)^{1/2} \simeq 10^{-35}$m that arises naturally in quantum theories of gravity).

However, not withstanding such caveats, it remains the case that one of Einstein's major contributions to physics was his realisation this it is possible to generalise the mathematical model of spacetime whilst keeping the basic ideas of (i) being able to locate a spacetime point via the values of a set of real numbers; and (ii) being able to describe the dynamical evolution of a system using differential equations. Specifically, in general relativity a spacetime is modelled by a 'differentiable manifold', of which the Euclidean space $\mathbb{R}^4$ of special relativity is just a special example. This remains one of the major motivations for studying differential geometry. However, as emphasised earlier, differential geometry enters into many other areas of modern theoretical physics, and it has become an indispensable tool for many scientists who work in these fields.

## 2.2 The Main Definitions

### 2.2.1 Coordinate charts

We must proceed now to give the formal definition of a differentiable manifold. For convenience, and unless stated otherwise, it will be assumed that $\mathcal{M}$ is a connected[1], Hausdorff topological space.

**Definition 2.1**

1. An $m$-dimensional ($m < \infty$) *coordinate chart* on a topological space $\mathcal{M}$ is a pair $(U, \phi)$ where $U$ is an open subset of $\mathcal{M}$ (called the *domain* of the coordinate chart) and $\phi : U \to \mathbb{R}^m$ is a homeomorphism of $U$ onto an open subset of the Euclidean space $\mathbb{R}^m$ equipped with its usual metric topology. This is illustrated in Figure 2.1.

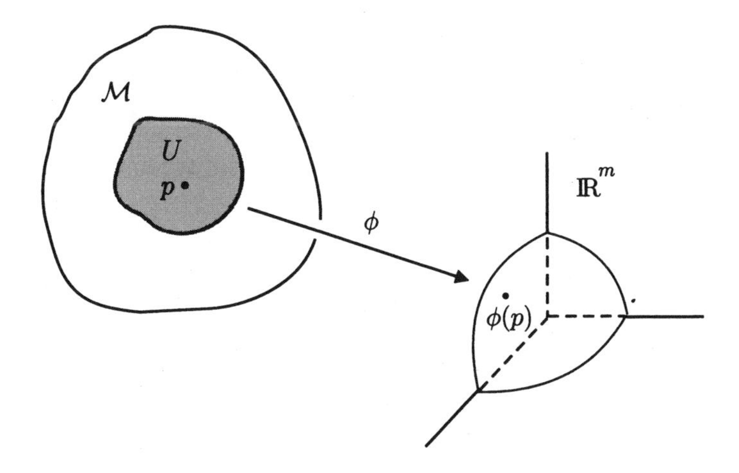


Figure 2.1: A local coordinate chart.

   If $U = \mathcal{M}$ then the coordinate chart is said to be *globally defined*; otherwise it is *locally defined*.

2. Let $(U_1, \phi_1)$ and $(U_2, \phi_2)$ be a pair of $m$-dimensional coordinate charts with $U_1 \cap U_2 \neq \emptyset$. Then the *overlap function* between the two coordinate charts is the map $\phi_2 \circ \phi_1^{-1}$ from the open subset $\phi_1(U_1 \cap U_2) \subset \mathbb{R}^m$ onto the open subset $\phi_2(U_1 \cap U_2) \subset \mathbb{R}^m$ (see Figure 2.2).

[1]A topological space is said to be *connected* if it cannot be written as the union of two disjoint open sets. Any topological space $X$ that is not connected can be decomposed uniquely into a union of disjoint open subsets, know as the *components* of $X$.

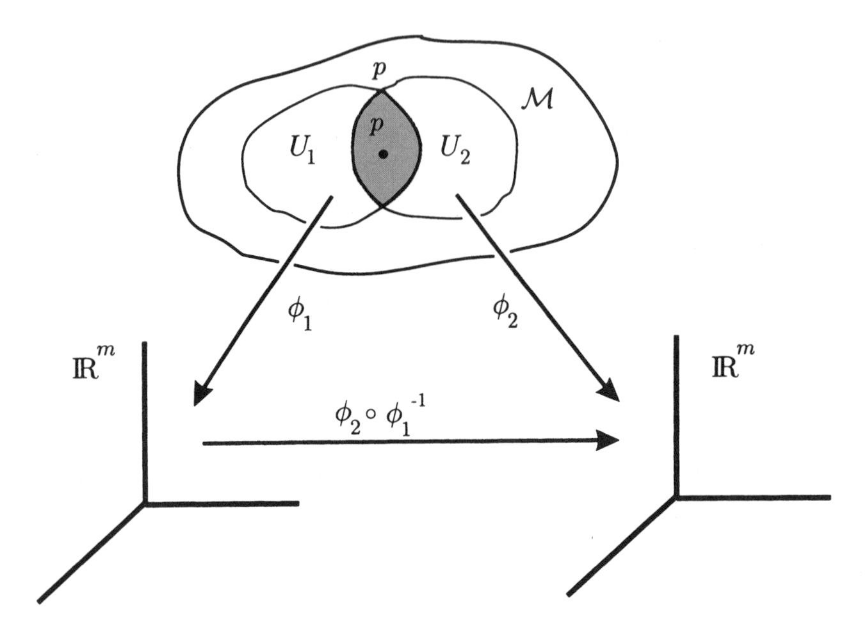


Figure 2.2: The overlap function of two coordinate charts.

3. An *atlas of dimension m* on $\mathcal{M}$ is a family of $m$-dimensional coordinate charts $(U_i, \phi_i)_{i\in I}$ (where $I$ is an index set) such that

   (a) $\mathcal{M}$ is covered by the family in the sense that $\mathcal{M} = \bigcup_{i\in I} U_i$;

   (b) each overlap function $\phi_j \circ \phi_i^{-1}$, $i, j \in I$, is a $C^\infty$ map[2] from $\phi_i(U_i \cap U_j)$ to $\phi_j(U_i \cap U_j)$ in $\mathbb{R}^m$.

   An atlas is said to be *complete* if it is maximal—*i.e.*, it is not contained in any other atlas.

   For a complete atlas, the family $(U_i, \phi_i)_{i\in I}$ is called a *differential structure* on $\mathcal{M}$ of dimension $m$. The topological space $\mathcal{M}$ is then said to be a *differentiable manifold*; or an *m-manifold* if it is useful to indicate the dimension explicitly.

---

[2]A function $f : \mathbb{R}^n \to \mathbb{R}^m$ is said to be $C^k$ if all its partial derivatives up to, and including, order $k$ exist, and if the derivatives of order $k$ are all continuous. A $C^\infty$ *function* is one for which all partial derivatives of all orders exist. These definitions extend immediately to a function that is defined only on some open subset of Euclidean space.

4. A point $p \in U \subset \mathcal{M}$, has the *coordinates* $(\phi^1(p), \phi^2(p), \ldots \phi^m(p)) \in \mathbb{R}^m$ with respect to the chart $(U, \phi)$, where the *coordinate functions* $\phi^\mu : U \to \mathbb{R}$, $\mu = 1, 2, \ldots, m$, are defined in terms of the projection functions $u^\mu : \mathbb{R}^m \to \mathbb{R}$, $u^\mu(x) := x^\mu$, as

$$\phi^\mu(p) := u^\mu(\phi(p)). \tag{2.2.1}$$

The coordinate functions are often written as $x^\mu$, $\mu = 1, 2, \ldots, m$, and the coordinates of a particular point $p$ are written as the $m$-tuple of real numbers $(x^1(p), x^2(p), \ldots, x^m(p))$.

**Comments**

1. As we shall see, the fact that the coordinates $x^\mu$ are to be thought of as (local) *functions* on the manifold plays a key role in the modern development of differential geometry.

2. There are analogous definitions of $C^k$, $(k < \infty)$ and $C^\omega$ manifolds in which the overlap functions are required to be $k$-times differentiable, and real analytic, respectively.

3. Similarly, there is the concept of a *complex* manifold in which the Euclidean space $\mathbb{R}^m$ is replaced with $\mathbb{C}^m$, and the overlap functions are required to be holomorphic. This latter requirement introduces radical changes in the idea of a manifold and requires a separate study.

4. Many of the ideas of differential geometry can be extended to the situation where the dimension is *infinite*, but considerable care is needed. This arises because on any (real) finite-dimensional vector space $V$ there is a unique Hausdorff topology with respect to which the vector space operations of vector addition, and scalar multiplication, are continuous (*i.e.*, such that $V$ is a *topological vector space*)[3]. However, there exist many different types of infinite-dimensional topological vector space, each of which is potentially associated with a corresponding type of infinite-dimensional manifold in which the vector space is the range space of the coordinate-chart functions $\phi : U \to V$.

[3]This is just the standard topology on the vector space $\mathbb{R}^n$ to which, of course, $V$ is isomorphic if $n = \dim V$. For a proof of this and related theorems on topological vector spaces see Treves (1967).

Infinite-dimensional differential geometry arises in several places in modern theoretical physics, usually in the context of spaces of non-linear functions: for example, the non-linear $\sigma$-model; or the space of embeddings of a 3-manifold in a 4-manifold. The situation that is closest to the finite-dimensional cases is when the coordinate spaces are chosen to be Banach spaces; a classic reference on this subject is Lang (1972). It would be beyond the scope of the present book to deal properly with infinite-dimensional manifolds, but at various points in the text I will indicate briefly whether or not the particular topic under discussion can be extended to the infinite-dimensional case.

5. If $\mathcal{M}_1$ and $\mathcal{M}_2$ are two differentiable manifolds then the Cartesian product $\mathcal{M}_1 \times \mathcal{M}_2$ can be given a manifold structure in a natural way [Exercise!]. The dimension of $\mathcal{M}_1 \times \mathcal{M}_2$ is the sum of the dimensions of $\mathcal{M}_1$ and $\mathcal{M}_2$. □

It is most unusual in practice to show that a given space is a manifold by directly examining the differentiability properties of overlap functions associated with various open coverings of the space. A more typical procedure is to start with a few very special spaces that can trivially be seen to be manifolds (such as the Euclidean space $\mathbb{R}^n$—see below) and then to construct a subspace of one such manifold in a way that is guaranteed to give the subspace a differential structure. This involves the important idea of a 'submanifold' of a given manifold.

**Definition 2.2**

A subset $\mathcal{N}$ of a differentiable manifold $\mathcal{M}$ is a $C^\infty$-*submanifold* of $\mathcal{M}$ if every point of $\mathcal{N}$ lies in some chart $(U, \phi)$ with

$$\phi(\mathcal{N} \cap U) = \phi(U) \cap \mathbb{R}^k \text{ where } 0 < k \leq m. \tag{2.2.2}$$

This is sketched in Figure 2.3.

### 2.2.2 Some examples of differentiable manifolds

**1. The Euclidean space $\mathbb{R}^m$:** The Euclidean space $\mathbb{R}^m$ equipped with its usual, non-compact, metric topology, can be given a differ-

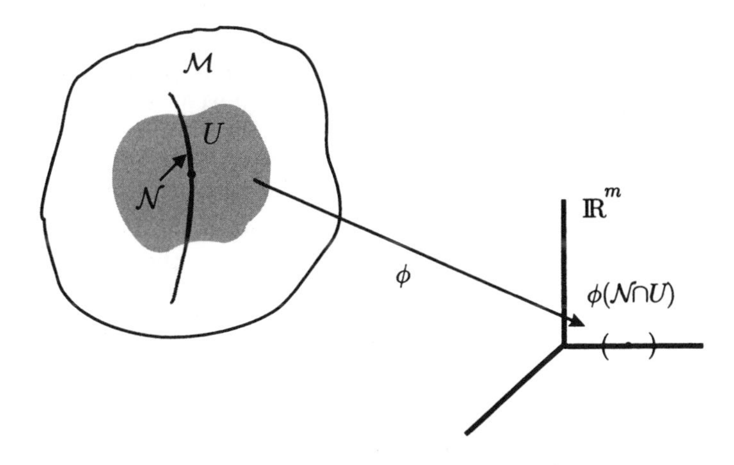


Figure 2.3: A submanifold $\mathcal{N}$ of a manifold $\mathcal{M}$.

ential structure with a globally-defined coordinate chart in which the coordinates of a vector (*i.e.*, an $n$-tuple of real numbers) $\vec{x}$ are defined to be the components $(x^1, x^2, \ldots, x^m)$ (*i.e.*, the elements of the $n$-tuple).

More generally, any finite-dimensional vector space $V$ can be regarded as a differentiable manifold by choosing any basis set of vectors for $V$ and using this to map $V$ isomorphically onto $\mathbb{R}^m$ in the usual way[4], where $m = \dim V$.

Note that any open subset $W$ of $\mathbb{R}^m$ inherits an $m$-dimensional differential structure from $\mathbb{R}^m$ in which the coordinate charts have domains that are the intersection of appropriate open subsets of $W$ with the domain ($\mathbb{R}^m$) of the globally-defined chart on $\mathbb{R}^m$. However, $W$ may be very non-trivial topologically even though its 'carrier' space $\mathbb{R}^m$ is trivial in this respect (more precisely, $\mathbb{R}^m$ is a *contractible* topological space).

**2. The circle $S^1$:** The circle $S^1$ can be thought of as the subset $\{(x, y) \in \mathbb{R}^2 \mid x^2 + y^2 = 1\}$ of the Euclidean space $\mathbb{R}^2$. If the latter

[4]If $\{e_1, e_2, \ldots, e_m\}$ is a basis for $V$, the vector $v = \sum_{i=1}^m v^i e_i$ is mapped to the $n$-tuple $(v^1, v^2, \ldots, v^m) \in \mathbb{R}^m$.

is equipped with its usual metric topology, then $S^1$ is clearly a closed and bounded subset, and hence—by the Heine-Borel theorem—its subspace topology is *compact* (see Section 1.4.8).[5]

In trying to give $S^1$ a differential structure it might seem natural to invoke the familiar angular coordinate $\theta$. However, this is not defined *globally* on the circle in the sense that it is not a continuous function at the point where $\theta = 0$ and $\theta = 2\pi$. In fact, there is no way of parameterising the circle globally with a *single* coordinate function.

This situation is generally true: *i.e.*, it is not possible to locate a point anywhere on a typical $m$-dimensional manifold with just a single coordinate chart. Of course, one example where this *is* possible is the Euclidean space $\mathbb{R}^m$; more generally, any $m$-dimensional manifold $\mathcal{M}$ for which there is a single globally-defined coordinate chart is necessarily homeomorphic to $\mathbb{R}^m$.

In the case of the circle $S^1$, one possibility would be to use a pair of overlapping angular coordinates. Another explicit set of possible coordinate charts is shown in Figure 2.4, where

$$\begin{aligned} U_1 &:= \{(x,y) \mid x > 0\} & \phi_1(x,y) &:= y \\ U_2 &:= \{(x,y) \mid x < 0\} & \phi_2(x,y) &:= y \\ U_3 &:= \{(x,y) \mid y > 0\} & \phi_3(x,y) &:= x \\ U_4 &:= \{(x,y) \mid y < 0\} & \phi_4(x,y) &:= x \end{aligned}$$

Note that although the coordinate functions have been written as functions of both $x$ and $y$ it is understood that these variables are subject to the constraint $x^2 + y^2 = 1$ and that $(x, y)$ means the point on the *circle* with this (constrained) value of $x$ and $y$—*i.e.*, the circle is a space of dimension one, not two.

To see that the overlap functions are differentiable consider, for example, the overlap of $U_1$ and $U_3$. In $U_1 \cap U_3$ we have

$$y = (1 - x^2)^{\frac{1}{2}} \text{ with } 0 < y < 1 \text{ and } 0 < x < 1. \tag{2.2.3}$$

[5]Note that referring to the subset $\{(x, y) \in \mathbb{R}^2 \mid x^2 + y^2 = 1\}$ as 'the' circle, involves a slight abuse of language in the sense that any other subset of $\mathbb{R}^2$ that is homeomorphic to this particular subset would be just as good a choice from a topological perspective.

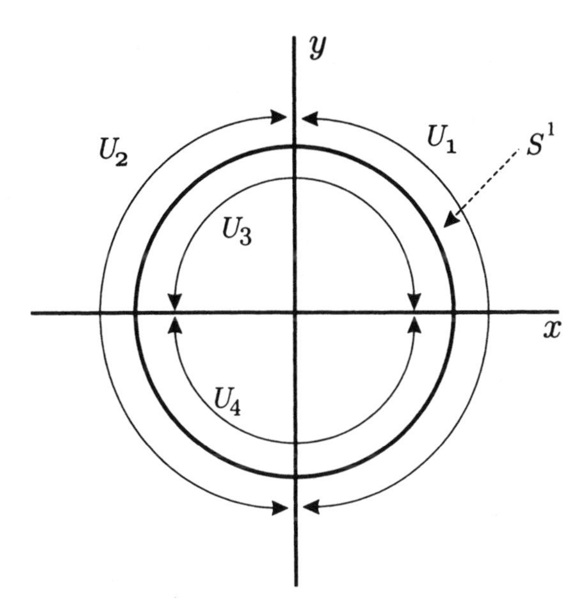


Figure 2.4: An atlas for the circle $S^1$.

Thus $\phi_3(x)^{-1} = (x, (1 - x^2)^{\frac{1}{2}})$, and so

$$\phi_1 \circ \phi_3^{-1}(x) = (1 - x^2)^{\frac{1}{2}} \tag{2.2.4}$$

which is indeed infinitely differentiable for this range of values for $x$ and $y$.

Note that if $\mathbb{R}^2$ is regarded as a manifold, then the subset $S^1$ is a compact, one-dimensional submanifold.

**3. The torus $T^2$:** The *torus* is defined to be the Cartesian product $S^1 \times S^1$ of two circles. As the product of two compact topological spaces, $T^2$ is itself compact (see Section 1.4.8) and can be equipped with the product of the differential structures on each individual $S^1$. This compact, two-dimensional manifold can be parameterised locally by specifying the values of two angles.

**4. The $n$-sphere:** The $n$-sphere can be regarded as the subset $S^n := \{\vec{x} \in \mathbb{R}^{n+1} \mid \vec{x} \cdot \vec{x} = 1\}$, of the Euclidean space $\mathbb{R}^{n+1}$; as such it acquires a topology which—by the Heine-Borel theorem—is compact. It can be given an explicit differential structure by means of stereographic projection from the north and south poles (the maps $\phi_1$ and $\phi_2$ respectively) as illustrated in Figure 2.5. Detailed consideration

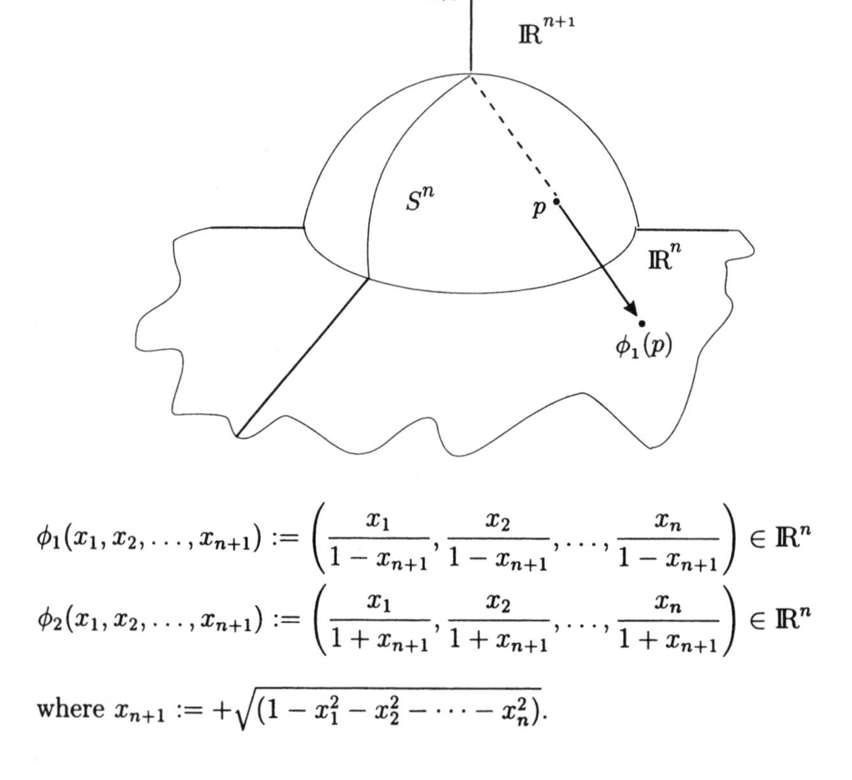


$$\phi_1(x_1, x_2, \ldots, x_{n+1}) := \left(\frac{x_1}{1 - x_{n+1}}, \frac{x_2}{1 - x_{n+1}}, \ldots, \frac{x_n}{1 - x_{n+1}}\right) \in \mathbb{R}^n$$

$$\phi_2(x_1, x_2, \ldots, x_{n+1}) := \left(\frac{x_1}{1 + x_{n+1}}, \frac{x_2}{1 + x_{n+1}}, \ldots, \frac{x_n}{1 + x_{n+1}}\right) \in \mathbb{R}^n$$

where $x_{n+1} := +\sqrt{(1 - x_1^2 - x_2^2 - \cdots - x_n^2)}$.

Figure 2.5: Stereographic projection from an $n$-sphere.

of the overlap functions shows that this gives $S^n$ the structure of a differentiable manifold and that it is a submanifold of $\mathbb{R}^{n+1}$.

### 2.2.3 Differentiable maps

A very important general idea in mathematics is that of a structure-preserving map between two sets that are equipped with the same type of mathematical structure. For example, in group theory this would be a homomorphism; in topology, a structure-preserving map is a continuous map between two topological spaces.

In differential geometry, the role of a structure-preserving map is played by a '$C^r$-function' between two manifolds, which is defined as follows:

**Definition 2.3**

1. The *local representative* of a function $f$ (from a manifold $\mathcal{M}$ to a manifold $\mathcal{N}$) with respect to the coordinate charts $(U, \phi)$ and $(V, \psi)$ on $\mathcal{M}$ and $\mathcal{N}$ respectively, is the map

$$\psi \circ f \circ \phi^{-1} : \phi(U) \subset \mathbb{R}^m \to \mathbb{R}^n \tag{2.2.5}$$

that is illustrated in Figure 2.6.

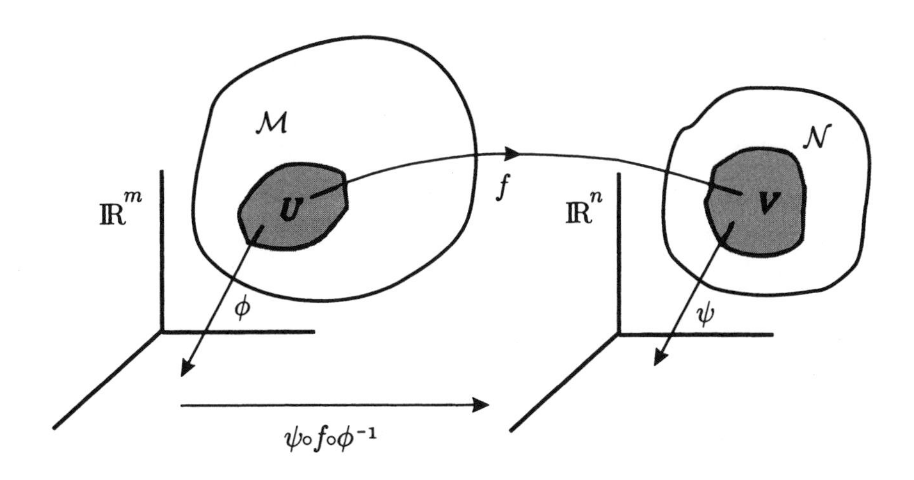


Figure 2.6: A local representative of a function $f : \mathcal{M} \to \mathcal{N}$.

2. A map $f : \mathcal{M} \to \mathcal{N}$ is a *$C^r$-function* if, for all coverings of $\mathcal{M}$ and $\mathcal{N}$ by coordinate neighbourhoods, the local representatives are $C^r$ functions as defined in the standard real analysis of functions between the topological vector spaces $\mathbb{R}^m$ and $\mathbb{R}^n$.

In particular, a *differentiable function* is defined to be a $C^1$ function.

A function that is $C^\infty$ is also said to be *smooth*.

3. A function $f : \mathcal{M} \to \mathcal{N}$ is a $C^r$*-diffeomorphism* if $f$ is a bijection with the property that both $f$ and its inverse are $C^r$ functions.

**Comments**

1. In trying to show that any given map $f : \mathcal{M} \to \mathcal{N}$ is $C^r$, it suffices (fortunately!) to show that the local representative of $f$ is $C^r$ for a *single* covering of $\mathcal{M}$ and $\mathcal{N}$ by coordinate charts. The $C^r$ nature of the local representative with respect to any other coverings by coordinate charts then follows from the differentiability requirement on the overlap functions.

2. Two manifolds that are diffeomorphic (*i.e.*, there exists a diffeomorphism between them) are 'equivalent' in the sense that they can be regarded as two different concrete copies of a single abstract Platonic manifold situated somewhere in the realms above. This is in the same sense that two groups that are isomorphic can be regarded as being the 'same' group.

3. The set of all diffeomorphisms of a manifold $\mathcal{M}$ onto itself forms a group Diff($\mathcal{M}$). Groups of this type play an important role in several branches of modern theoretical physics. □

## 2.3 Tangent Spaces

### 2.3.1 The intuitive idea

One of the most important ideas in differential geometry is that of a 'tangent space' to a manifold. This is based in part on the intuitive geometric idea of a tangent plane to a surface, as sketched in Figure 2.7. Thus the tangent space at the point $\vec{x} \in S^n$ looks as if it should be defined by

$$T_{\vec{x}}S^n := \{\vec{v} \in \mathbb{R}^{n+1} \mid \vec{x} \cdot \vec{v} = 0\}. \tag{2.3.1}$$

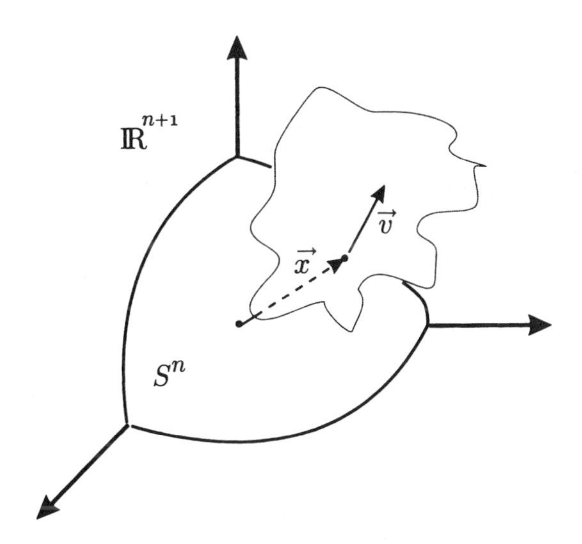


Figure 2.7: A tangent space for the sphere $S^n$ embedded in $\mathbb{R}^{n+1}$.

However, it transpires that tangent space structure is also deeply connected with the local differentiable properties of functions on the manifold, and this gives a more algebraic slant to the idea. In fact, there are several—superficially quite different—definitions of a tangent space, some of which emphasise the geometric picture and some the algebraic. As we shall see, for finite-dimensional manifolds these definitions are ultimately equivalent, and it is largely a matter of taste that determines the approach to be adopted in any particular case. The situation with regard to infinite-dimensional manifolds is, however, quite different and many of the finite-dimensional definitions are no longer appropriate. One of the main aims of this section is to present these different approaches to the definition of a tangent space and to show that they are indeed equivalent. Thus the discussion is firmly grounded in the world of finite-dimensional differential geometry. Nevertheless, the presentation will begin with the particular definition that *does* often generalise to an infinite-dimensional manifold.

This particular approach is slanted heavily towards the geometric view of a tangent space as being 'tangent' to the manifold, rather as in the picture above for the case of a sphere. However, the problem with this picture is that it is just that—a picture. In particular, it involves embedding the sphere in a particular vector space $\mathbb{R}^{n+1}$, and then regarding a tangent space as a specific linear subspace of this vector space. But one of the main aims of modern differential geometry is to present ideas in a way that is intrinsic to the manifold itself and, in particular, is not dependent on an embedding in some (largely arbitrary) higher-dimensional vector space. For example, the definition of a differentiable manifold itself made no reference to such a space, and neither should the definition of a tangent space.

The critical question, therefore, is to understand what should replace the intuitive idea of a tangent vector as something that is tangent to a surface in the usual sense and which, in particular, 'sticks out' into the containing space. The answer is that a tangent vector is to be understood as something that is tangent to a *curve* in the manifold. In Figure 2.7 it is clear that a curve can be drawn on the sphere with the property that, at the point $\vec{x} \in S^n$, its tangent vector (in the usual sense of a curve in a Euclidean space) is equal to the vector $\vec{v}$. The crucial point here is that the curve itself lies in the manifold $S^n$, *not* in the surrounding $\mathbb{R}^{n+1}$, and therefore this idea of a tangent vector as a 'tangent to a curve' can be generalised to an arbitrary manifold without it first needing to be embedded in a higher-dimensional vector space. Thus the general idea is to *define* a tangent vector in terms of a curve to which—pictorially—it might be deemed to be tangent.

However, things are not quite as simple as this because, for example, in the case above of $S^n$—which *is* embedded in a convenient space—it is clear that there are many curves on the manifold that are 'tangent' to the given vector $\vec{v}$ (see Figure 2.8). This leads to the idea of a tangent vector as an appropriate *equivalence class* of curves, and this is the definition that we shall now present.

### 2.3.2 A tangent vector as an equivalence class of curves

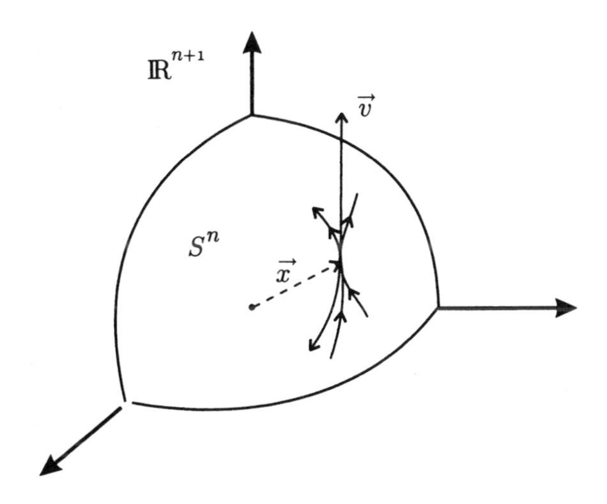


Figure 2.8: Many curves are tangent to the same vector $\vec{v}$.

**Definition 2.4**

1. A *curve* on a manifold $\mathcal{M}$ is a smooth (*i.e.*, $C^\infty$) map $\sigma$ from some interval $(-\epsilon, \epsilon)$ of the real line into $\mathcal{M}$.

   Note that the 'curve' is defined to be the map itself, *not* the set of image points in $\mathcal{M}$; this is illustrated in Figure 2.9. It is important to remember this distinction between a function and its set of image points.

2. Two curves $\sigma_1$ and $\sigma_2$ are *tangent* at a point $p$ in $\mathcal{M}$ if

   (a) $\sigma_1(0) = \sigma_2(0) = p$;

   (b) in some local coordinate system $(x^1, x^2, \ldots, x^m)$ around the point, the two curves are 'tangent' in the usual sense as curves in $\mathbb{R}^m$:

   $$\frac{dx^i}{dt}(\sigma_1(t))\bigg|_{t=0} = \frac{dx^i}{dt}(\sigma_2(t))\bigg|_{t=0} \tag{2.3.2}$$

   for $i = 1, 2, \ldots, m$.

   Note that if $\sigma_1$ and $\sigma_2$ are tangent in one coordinate system, then they are tangent in any other coordinate system that covers the point $p \in \mathcal{M}$. Thus the definition is an intrinsic one, *i.e.*, it is independent of coordinate system.

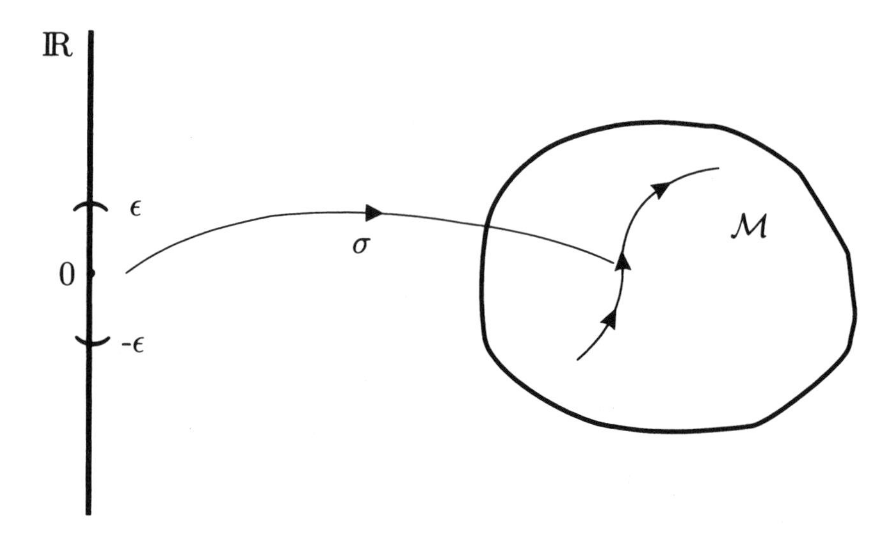


Figure 2.9: A curve is a function.

3. A *tangent vector* at $p \in \mathcal{M}$ is an equivalence class of curves in $\mathcal{M}$ where the equivalence relation between two curves is that they are tangent at the point $p$. The equivalence class of a particular curve $\sigma$ will be denoted $[\sigma]$.

4. The *tangent space* $T_p\mathcal{M}$ to $\mathcal{M}$ at a point $p \in \mathcal{M}$ is the set of all tangent vectors at the point $p$.

   The *tangent bundle* $T\mathcal{M}$ is defined as $T\mathcal{M} := \bigcup_{p\in\mathcal{M}} T_p\mathcal{M}$.

   There is a natural *projection map* $\pi : T\mathcal{M} \to \mathcal{M}$ that associates with each tangent vector the point $p$ in $\mathcal{M}$ at which it is tangent. The inverse image (the *fibre* over $p$) of any point $p$ under the map $\pi$ is thus the set of all vectors that are tangent to the manifold at that point—*i.e.*, the tangent space $T_p\mathcal{M}$ (see Figure 2.10). This is a special case of the general idea of a fibre bundle—something to which we shall return later.

### Comments

1. This definition of a tangent vector is consistent with the intuitive geometrical picture when—as in the case of a sphere embedded

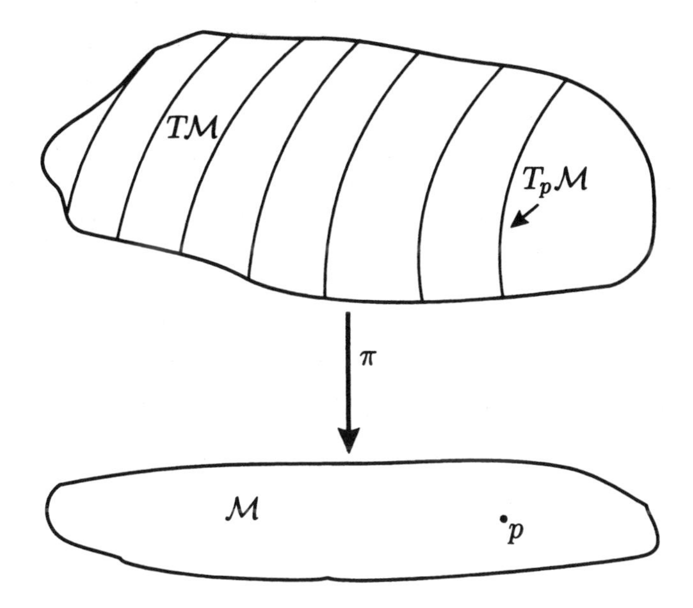


Figure 2.10: The tangent bundle $T\mathcal{M}$.

in a Euclidean space—there is a natural way of representing tangent vectors as elements of the containing vector space.

This remark also extends to the tangent bundle $T\mathcal{M}$ which, in the case of a sphere, can be represented as:

$$TS^n = \{\, (\vec{x}, \vec{v}) \in \mathbb{R}^{n+1} \times \mathbb{R}^{n+1} \mid \vec{x} \cdot \vec{x} = 1 \text{ and } \vec{x} \cdot \vec{v} = 0 \,\}. \qquad (2.3.3)$$

2. A tangent vector $v$ in $T_p\mathcal{M}$ can be used as a 'directional derivative' on functions $f$ on $\mathcal{M}$ by defining:

$$v(f) := \left. \frac{df(\sigma(t))}{dt} \right|_{t=0} \qquad (2.3.4)$$

where $\sigma$ is any curve in the equivalence class represented by $v$, *i.e.*, $v = [\sigma]$. (Exercise: Show that this definition does not depend on the particular choice of $\sigma$ in the equivalence class $v$.) The idea that a tangent vector can be regarded as a type of differential operator is a crucial ingredient in the other—more algebraic—definitions of tangent space structure. We shall return to this point later. □

### 2.3.3 The vector space structure on $T_p\mathcal{M}$

In the heuristic, pictorial representation of tangent vectors to, for example, $S^n$ embedded in $\mathbb{R}^{n+1}$, it is clear that (i) any two tangent vectors at a point $p$ can be added to give a third; and (ii) a real multiple of a tangent vector is itself a tangent vector. This suggests strongly that it should be possible to make $T_p\mathcal{M}$ into a vector space within the framework of the general definition of a tangent vector given above.

**Theorem 2.1** *The tangent space $T_p\mathcal{M}$ carries a structure of a real vector space.*

**Proof**

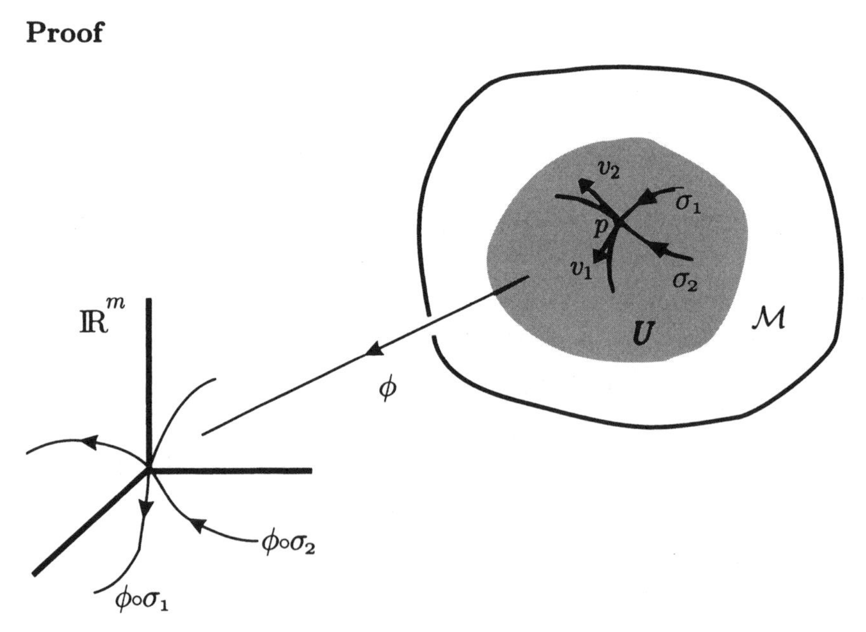


Figure 2.11: The addition of two equivalence classes of curves.

In Figure 2.11, $\sigma_1$ and $\sigma_2$ are representative curves for the two tangent vectors $v_1$ and $v_2$ in $T_p\mathcal{M}$. We use a coordinate chart $(U, \phi)$ around $p \in \mathcal{M}$ with the property that $\phi(p) = \vec{0} \in \mathbb{R}^m$, and then consider the image curves $\phi \circ \sigma_1$ and $\phi \circ \sigma_2$ which map an open interval into $\mathbb{R}^m$.

Of course, the curves $\sigma_1$ and $\sigma_2$ cannot be 'added' directly since the set $\mathcal{M}$ in which they take their values is not a vector space. However, the local coordinate space $\mathbb{R}^m$ *is* a vector space, and hence it is legitimate to consider the sum $t \mapsto \phi \circ \sigma_1(t) + \phi \circ \sigma_2(t)$ which is a curve in $\mathbb{R}^m$ and which, like both $\sigma_1$ and $\sigma_2$, passes through the null vector $\vec{0}$ when $t = 0$. It follows that the map

$$t \mapsto \phi^{-1} \circ (\phi \circ \sigma_1(t) + \phi \circ \sigma_2(t)) \tag{2.3.5}$$

is a curve in $\mathcal{M}$ that passes through the point $p$ when $t = 0$. Then define

$$\text{(i) } v_1 + v_2 := [\phi^{-1} \circ (\phi \circ \sigma_1 + \phi \circ \sigma_2)] \tag{2.3.6}$$

$$\text{(ii) } rv := [\phi^{-1} \circ (r\phi \circ \sigma)] \text{ for all } r \in \mathbb{R}. \tag{2.3.7}$$

It can be shown [Exercise!] that

(a) these definitions are independent of the choice of chart $(U, \phi)$ and representatives $\sigma_1$ and $\sigma_2$ of the tangent vectors $v_1$ and $v_2$;

(b) under these definitions, $T_p\mathcal{M}$ is a real vector space.

Note that this means that $T\mathcal{M}$ is a *vector bundle.* More details of this will be given later.

### 2.3.4 The push-forward of an equivalence class of curves

To some extent, a tangent space can be regarded as a local 'linearization' of the manifold, and it is of considerable importance that a map $h$ between two manifolds $\mathcal{M}$ and $\mathcal{N}$ can be 'linearized' with the aid of the tangent spaces and the vector space structure that they carry. This is the concept of the 'push-forward' map $h_*$ that maps the tangent spaces of $\mathcal{M}$ *linearly* into the tangent spaces of $\mathcal{N}$ according to the following definition.

**Definition 2.5**

If $h : \mathcal{M} \to \mathcal{N}$ and $v \in T_p\mathcal{M}$ then the '*push-forward*' $h_*(v)$ in $T_{h(p)}\mathcal{N}$ is defined by (see Figure 2.12)

$$h_*(v) := [h \circ \sigma] \text{ where } v = [\sigma]. \tag{2.3.8}$$

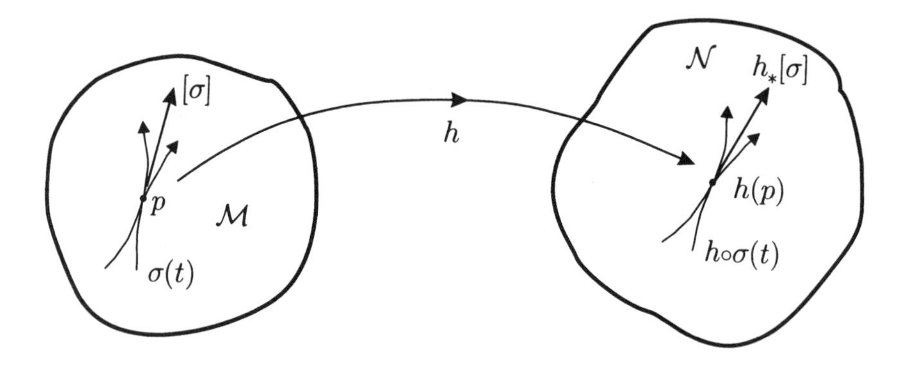


Figure 2.12: The push-forward operation on tangent vectors.

**Exercises**

1. Show that this definition is independent of the particular choice of the curve $\sigma$ in the equivalence class $v \in T_p\mathcal{M}$.

2. Show that the map $h_*$ is *linear*. That is, for all $v_1, v_2 \in T_p\mathcal{M}$ and $r \in \mathbb{R}$,

$$h_*(v_1 + v_2) = h_*(v_1) + h_*(v_2) \tag{2.3.9}$$
$$h_*(rv) = rh_*(v). \tag{2.3.10}$$

3. If $\mathcal{M}$, $\mathcal{N}$ and $\mathcal{P}$ are manifolds, and if $h : \mathcal{M} \to \mathcal{N}$ and $k : \mathcal{N} \to \mathcal{P}$ show that

$$(k \circ h)_* = k_* h_* \tag{2.3.11}$$

which relates to the diagrams

$$\begin{array}{ccccc} T\mathcal{M} & \xrightarrow{h_*} & T\mathcal{N} & \xrightarrow{k_*} & T\mathcal{P} \\ \downarrow & & \downarrow & & \downarrow \\ \mathcal{M} & \xrightarrow{h} & \mathcal{N} & \xrightarrow{k} & \mathcal{P} \end{array} \quad \text{and} \quad \begin{array}{ccc} T\mathcal{M} & \xrightarrow{(k\circ h)_*} & T\mathcal{P} \\ \downarrow & & \downarrow \\ \mathcal{M} & \xrightarrow{k\circ h} & \mathcal{P} \end{array} \tag{2.3.12}$$

In effect, Eq. (2.3.11) says that if $g : \mathcal{M} \to \mathcal{P}$ is a map from a manifold $\mathcal{M}$ to a manifold $\mathcal{P}$, and if $v$ is a tangent vector at $p \in \mathcal{M}$, then the push-forward $g_*(v) \in T_{g(p)}\mathcal{P}$ of $v$ is independent of factorising $g$ through some intermediate manifold $\mathcal{N}$ as $\mathcal{M} \xrightarrow{h} \mathcal{N} \xrightarrow{k} \mathcal{P}$ with $g = k \circ h$. This 'functorial'[6] property is necessary for the idea of the push-forward operation to be of much use: otherwise it would be as if sending a letter from London to New York by air produces a different outcome if the plane stops to refuel somewhere along the way!

### 2.3.5 Tangent vectors as derivations

So far, our description of a tangent vector has been (hopefully!) geometrically appealing but, admittedly, rather abstract. In particular, a reader who has already encountered the coordinate-based approach to differential geometry (for example, by taking an introductory course on general relativity) may be wondering if, or how, the definition of a tangent vector given above can be related to the familiar 'object with a superscript index' $\xi^\mu$ much favoured in such approaches.

A powerful way of uncovering this relation is to discuss first an alternative rigorous approach to the idea of a tangent vector that is essentially *algebraic* in form, rather than geometric, and which is based on the idea of a 'directional derivative'. The ensuing interplay between the geometric and the algebraic ways of understanding tangent vectors is one of the central features of modern differential geometry.

Recall that the directional derivative of a function $f$ along a tangent vector $v$ is defined as (Eq. (2.3.4))

$$v(f) := \left.\frac{df(\sigma(t))}{dt}\right|_{t=0} \quad \text{where } [\sigma] = v \tag{2.3.13}$$

which enables the equivalence class of curves $v \in T_p\mathcal{M}$ to act as a type of differential operator on the space $C^\infty(\mathcal{M})$ of real-valued

[6]The natural language in which to describe this type of property is 'category theory', and a *functor* is a structure-preserving operation between categories. I will not introduce such ideas formally in this book, although they do lie behind many of the constructions made in differential geometry. A good introduction to a category-based view of differential geometry is given in Lang (1972).

differentiable functions on $\mathcal{M}$. The first 'algebraic' way of defining tangent vectors is to turn Eq. (2.3.13) around and *define* $T_p\mathcal{M}$ as a space of directional derivatives. This idea is developed by thinking of $C^\infty(\mathcal{M})$ as a ring over $\mathbb{R}$, and then regarding a tangent vector as a derivation map from this ring into $\mathbb{R}$. More precisely:

**Definition 2.6**

1. A *derivation* at a point $p \in \mathcal{M}$ is a map $v : C^\infty(\mathcal{M}) \to \mathbb{R}$ such that

$$\text{(i) } v(f+g) = v(f) + v(g) \text{ for all } f, g \in C^\infty(\mathcal{M}) \tag{2.3.14}$$
$$v(rf) = rv(f) \text{ for all } f \in C^\infty(\mathcal{M}) \text{ and } r \in \mathbb{R}; \tag{2.3.15}$$
$$\text{(ii) } v(fg) = f(p)v(g) + g(p)v(f) \text{ for all } f, g \in C^\infty(\mathcal{M}). \tag{2.3.16}$$

2. The set of all derivations at $p \in \mathcal{M}$ is denoted $D_p\mathcal{M}$.

**Comments**

1. Equations (2.3.14–2.3.15) merely assert that $v$ is a linear map from the vector space $C^\infty(\mathcal{M})$ to $\mathbb{R}$. The term 'derivation' is used because of the crucial 'Leibniz rule' in Eq. (2.3.16) which is clearly analogous to the rule in elementary calculus for taking the derivative at a fixed point of the product of two functions.

2. The space of derivations $D_p\mathcal{M}$ can readily be given the structure of a real vector space by defining

$$(v_1 + v_2)(f) := v_1(f) + v_2(f) \tag{2.3.17}$$
$$(rv)(f) := rv(f) \tag{2.3.18}$$

for all $v_1, v_2, v$ in $D_p(\mathcal{M})$, $f \in C^\infty(\mathcal{M})$ and $r \in \mathbb{R}$.

3. Note how much more natural (and easier!) is this definition of linear structure on $D_p\mathcal{M}$ as compared to the construction in Theorem 2.1 of the vector-space structure on $T_p\mathcal{M}$. This illustrates an obvious, but important, general rule: geometric ideas appear most naturally in the geometric approach to tangent vectors, and algebraic ideas appear most naturally in the algebraic approach.

Equation (2.3.13) defines a map from $T_p\mathcal{M}$ to $D_p\mathcal{M}$ that is clearly linear, and we shall show in Theorem 2.2 that $T_p\mathcal{M}$ and $D_p\mathcal{M}$ are, in fact, isomorphic. It follows that we can adopt whichever approach we wish—geometric or algebraic—according to the type of problem under discussion. There are a number of explicit examples of this procedure throughout the book. Unfortunately, this equality of approach to tangent-space structure only holds in finite-dimensional differential geometry, since the algebraic scheme is problematic in infinite dimensions. □

A very important example of a derivation is provided with the aid of any coordinate chart $(U, \phi)$ containing the point $p$ of interest (see Figure 2.13). Specifically, a set of derivations at $p$ is defined as[7]

$$\left(\frac{\partial}{\partial x^\mu}\right)_p f := \frac{\partial}{\partial u^\mu} f \circ \phi^{-1}\Big|_{\phi(p)} \; ; \quad \mu = 1, 2, \ldots, \dim \mathcal{M}. \qquad (2.3.19)$$

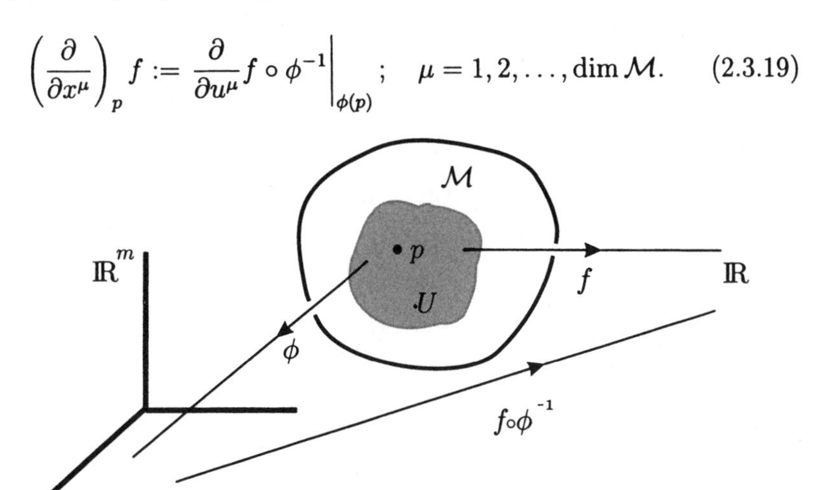


Figure 2.13: A local coordinate chart used to define a derivation.

[7]One should firmly resist any temptation to interpret the left hand side of Eq. (2.3.19) as meaning the 'partial-derivative' of $f$ evaluated at the point $p \in \mathcal{M}$, notwithstanding the alluring nature of the notation. The function $f$ is defined on the *manifold* $\mathcal{M}$, which generally does not possess the vector-space structure needed for a partial-derivative to be defined. Of course, partial-derivatives *can* be defined in the usual way for the local representatives $f \circ \phi^{-1} : \phi(U) \subset \mathbb{R}^n \to \mathbb{R}$ of $f$; and this is what is exploited in the definition Eq. (2.3.19), which—like all definitions—means no more and no less than it says.

We shall now show that these quantities form a *basis* set for the vector space $D_p\mathcal{M}$.

**Lemma** [8]

Let $(U, \phi)$ be a coordinate chart around $p \in \mathcal{M}$ with associated coordinate functions $(x^1, x^2, \ldots, x^m)$ and such that $x^\mu(p) = 0$ for all $\mu = 1, 2, \ldots, m = \dim \mathcal{M}$. Then for any $f \in C^\infty(\mathcal{M})$ there exists $f_\mu \in C^\infty(\mathcal{M})$ such that

$$\text{(i)} \qquad f_\mu(p) = \left(\frac{\partial}{\partial x^\mu}\right)_p f \tag{2.3.20}$$

$$\text{(ii)} \qquad f(q) = f(p) + \sum_{\nu=1}^{m} x^\nu(q)\, f_\nu(q) \tag{2.3.21}$$

for all $q$ in some open neighbourhood of $p$. Note that Eq. (2.3.21) means that the function $f$ can be written locally as

$$f = k_{f(p)} + \sum_{\nu=1}^{m} x^\nu f_\nu, \tag{2.3.22}$$

where, for any $c \in \mathbb{R}$, $k_c : \mathcal{M} \to \mathbb{R}$ denotes the constant function whose value is everywhere $c$.

**Proof**

Let $F := f \circ \phi^{-1}$ defined on some open ball $B$ around $\vec{0} \in \mathbb{R}^m$. Then, for $\vec{a} \in B$,

$$\begin{aligned} F(a_1, a_2, \ldots, a_m) &\equiv F(a_1, a_2, \ldots, a_m) - F(a_1, a_2, \ldots, a_{m-1}, 0) \\ &+ F(a_1, a_2, \ldots, a_{m-1}, 0) - F(a_1, a_2, \ldots, a_{m-2}, 0, 0) \\ &+ \ldots \\ &+ \ldots \\ &+ F(a_1, 0, \ldots, 0) - F(0, 0, \ldots, 0) + F(0, 0, \ldots, 0). \end{aligned} \tag{2.3.23}$$

This identity can be rewritten as

$$F(a_1, a_2, \ldots, a_m) = F(0, 0, \ldots, 0) + \sum_{\mu=1}^{m} F(a_1, a_2, \ldots, ta_\mu, 0, \ldots, 0)\Big|_{t=0}^{t=1}$$

[8] I have taken this lemma from the elegant little book by Hicks (1965).

$$
\begin{aligned}
&= F(0,0,\ldots,0) + \\
&\quad \sum_{\mu=1}^{m} \int_0^1 dt\, \frac{\partial F}{\partial u^\mu}(a_1, a_2, \ldots, a_{\mu-1}, ta_\mu, 0, \ldots, 0) a_\mu \\
&= F(0,0,\ldots,0) + \sum_{\mu=1}^{m} a_\mu F_\mu(a_1, a_2, \ldots, a_m) \quad (2.3.24)
\end{aligned}
$$

where

$$F_\mu(\vec{a}) := \int_0^1 dt\, \frac{\partial F}{\partial u^\mu}(a_1, a_2, \ldots, a_{\mu-1}, ta_\mu, 0, \ldots, 0) \quad (2.3.25)$$

is a $C^\infty$-function on $B \subset \mathbb{R}^m$.

Now let $f_\mu := F_\mu \circ \phi$, which is defined initially on some open neighbourhood of $p \in \mathcal{M}$ but which can then be extended arbitrarily to the rest of $\mathcal{M}$. Then

$$
\begin{aligned}
f(q) &= (f \circ \phi^{-1})(\phi(q)) = F(x^1(q), \ldots, x^m(q)) \\
&= F(\vec{0}) + \sum_{\mu=1}^{m} x^\mu(q)\, F_\mu(x^1(q), \ldots, x^m(q)) \\
&= f(p) + \sum_{\mu=1}^{m} x^\mu(q)\, f_\mu(q) \quad (2.3.26)
\end{aligned}
$$

and hence the desired representation Eq. (2.3.22) certainly exists.

The next step is to find an explicit form for $f_\mu(p)$. To this end we apply the derivation $\left(\frac{\partial}{\partial x^\mu}\right)_p$ at $p$ to both sides of Eq. (2.3.22) to give

$$\left(\frac{\partial}{\partial x^\mu}\right)_p f = \left(\frac{\partial}{\partial x^\mu}\right)_p k_{f(p)} + \sum_{\nu=1}^{m} \left[ \left( \left(\frac{\partial}{\partial x^\mu}\right)_p x^\nu \right) f_\nu(p) + x^\nu(p) \left(\frac{\partial}{\partial x^\mu}\right)_p f_\nu \right]. \quad (2.3.27)$$

However, if $v \in D_p\mathcal{M}$, then, using the Leibniz property of the derivation $v$, we get[9], for any $c \in \mathbb{R}$,

$$
\begin{aligned}
v(k_c) = v(ck_1) &= cv(k_1) = cv(k_1 \times k_1) \\
&= c(k_1(p)v(k_1) + k_1(p)v(k_1)) \\
&= c(1 \times v(k_1) + 1 \times v(k_1)) \\
&= 2cv(k_1) = 2v(k_c) \quad (2.3.28)
\end{aligned}
$$

[9]This is one of the nicest uses I know of the true—but otherwise unremarkable—fact that $1 = 1 \times 1$ (or, more precisely, that $k_1 = k_1 \times k_1$).

and hence $v(k_c) = 0$. Applying this result to Eq. (2.3.27) (and remembering that $x^\nu(p) = 0$) gives

$$\left(\frac{\partial}{\partial x^\mu}\right)_p f = \sum_{\nu=1}^{m} \left(\left(\frac{\partial}{\partial x^\mu}\right)_p x^\nu\right) f_\nu(p) = f_\mu(p) \tag{2.3.29}$$

which, since $\left(\frac{\partial}{\partial x^\mu}\right)_p x^\nu = \delta^\nu_\mu$ [Exercise!], proves the lemma. **QED**

### Corollary

$$\boxed{\text{If } v \in D_p(\mathcal{M}) \text{ then } v = \sum_{\mu=1}^{m} v(x^\mu) \left(\frac{\partial}{\partial x^\mu}\right)_p} \tag{2.3.30}$$

### Proof

If $x^\mu(p) \neq 0$ for $\mu = 1, 2, \ldots, m$ then define new coordinate functions $y^\mu := x^\mu - k_{x^\mu(p)}$ and apply the Lemma using this coordinate system. Then gives the equation between functions

$$\begin{aligned} f(\,) &= k_{f(p)}(\,) + \sum_{\mu=1}^{m} y^\mu(\,)\, f_\mu(\,) \\ &= k_{f(p)}(\,) + \sum_{\mu=1}^{m} \left(x^\mu(\,) - k_{x^\mu(p)}(\,)\right) f_\mu(\,). \end{aligned} \tag{2.3.31}$$

Applying the derivation $v$ to both sides of this equation—and remembering that $v$ vanishes on the constant functions $q \mapsto f(p)$ and $q \mapsto x^\mu(p)$—gives the result in Eq. (2.3.30). **QED**

### Comments

1. The set of real numbers $\{v^1, v^2, \ldots, v^m\}$, where $v^\mu := v(x^\mu)$, are called the *components* of the derivation $v \in D_p\mathcal{M}$ with respect to the given coordinate system. This is where 'an object with a superscript index' (*i.e.*, $v^\mu$) first arises in the modern approach to differential geometry: indeed, Eq. (2.3.30) is a key result in relating modern, 'coordinate-free' differential geometry to the more traditional, coordinate-based, approaches. We see here an example of the importance of the idea that the 'coordinates' $x^\mu$, $\mu = 1, 2, \ldots, m$ are

actually (local) *functions* on the manifold $\mathcal{M}$: the components $v^\mu$ of the derivation $v$ are obtained by letting $v$ act on these functions.

Note that the components of a given derivation will be different in different coordinate systems. We shall return to this later.

2. If $v = \sum_{\mu=1}^m v^\mu \left(\frac{\partial}{\partial x^\mu}\right)_p = 0$ for some set of real numbers $\{v^1, v^2, \ldots, v^m\}$ then, for all $\mu = 1, 2, \ldots, m$, we have $0 = v(x^\mu) = v^\mu$. This means that the set of vectors $\left(\frac{\partial}{\partial x^\mu}\right)_p$, $\mu = 1, 2, \ldots, m = \dim \mathcal{M}$ are linearly independent and span the vector space $D_p\mathcal{M}$. In particular, $\dim D_p\mathcal{M} = \dim \mathcal{M}$.

□

The result above can be used to prove the critical fact that $D_p\mathcal{M}$ and $T_p\mathcal{M}$ are isomorphic as vector spaces:

**Theorem 2.2** *The linear map $\iota : T_p\mathcal{M} \to D_p\mathcal{M}$ defined by (see Eq. (2.3.13))*

$$\iota(v)(f) := \left.\frac{df(\sigma(t))}{dt}\right|_{t=0} \quad \textit{where } [\sigma] = v \tag{2.3.32}$$

*is an isomorphism.*

### Proof

We know already that the map $\iota$ is linear, so we must show that (i) it is injective; and (ii) it is surjective.

**(i) $\iota$ is injective:** To show that $\iota$ is injective we note first that Eq. (2.3.32) is meaningful even if the function $f$ is defined only on some open neighbourhood of the point $\sigma(0) \in \mathcal{M}$. Now suppose that, for some $v_1, v_2 \in T_p\mathcal{M}$, we have $\iota(v_1) = \iota(v_2)$ with $v_1 = [\sigma_1]$ and $v_2 = [\sigma_2]$. Then consider Eq. (2.3.32) for $f$ ranging over the set of local coordinate functions $\{x^1, x^2, \ldots, x^m\}$. From the equality of $\iota(v_1)$ and $\iota(v_2)$ it follows that

$$\left.\frac{dx^\mu(\sigma_1(t))}{dt}\right|_{t=0} = \left.\frac{dx^\mu(\sigma_2(t))}{dt}\right|_{t=0} \tag{2.3.33}$$

for all $\mu = 1, 2, \ldots, m$. Thus the curves $\sigma_1$ and $\sigma_2$ are tangent at the point $p = \sigma_1(0) = \sigma_2(0)$, and hence $v_1 = [\sigma_1] = [\sigma_2] = v_2$; thus $\iota$ is injective.

**(ii) $\iota$ is surjective:** Let $v \in D_p\mathcal{M}$ and construct a curve $\sigma_v : (-\epsilon, \epsilon) \to \mathcal{M}$ such that:

$$\text{(i) } \sigma_v(0) = p; \tag{2.3.34}$$

$$\text{(ii) } v(x^\mu) = \frac{d}{dt} x^\mu \circ \sigma_v(t)\Big|_{t=0}. \tag{2.3.35}$$

This is a trivial task using a local coordinate system on $\mathcal{M}$. Then, if $f \in C^\infty(\mathcal{M})$, we have

$$v(f) = \sum_{\mu=1}^{m} v(x^\mu) \left(\frac{\partial}{\partial x^\mu}\right)_p f = \sum_{\mu=1}^{m} \frac{d}{dt} x^\mu \circ \sigma_v(t)\Big|_{t=0} \left(\frac{\partial}{\partial x^\mu}\right)_p f. \tag{2.3.36}$$

But

$$\begin{aligned} \frac{d}{dt} f \circ \sigma_v\Big|_{t=0} &= \frac{d}{dt}(f \circ \phi^{-1} \circ \phi \circ \sigma_v)\Big|_{t=0} \\ &= \sum_{\mu=1}^{m} \left(\frac{\partial}{\partial u^\mu}\right)(f \circ \phi^{-1})\Big|_{\phi(p)} \frac{d}{dt} u^\mu(\phi \circ \sigma_v)\Big|_{t=0} \\ &= \sum_{\mu=1}^{m} \left(\frac{\partial}{\partial x^\mu}\right)_p f \, \frac{d}{dt} x^\mu \circ \sigma_v(t)\Big|_{t=0} = v(f) \end{aligned} \tag{2.3.37}$$

which means that $[\sigma_v](f) = v(f)$. Hence $\iota : T_p\mathcal{M} \to D_p\mathcal{M}$ is surjective. **QED**

### Comments

1. Since $T_p\mathcal{M}$ and $D_p\mathcal{M}$ are isomorphic, it follows in particular that $\dim T_p\mathcal{M} = \dim \mathcal{M}$.

2. The isomorphism of $T_p\mathcal{M}$ and $D_p\mathcal{M}$ means that either can be used according to the situation under consideration. Any formula proved for tangent vectors defined as equivalence classes of curves in $T_p\mathcal{M}$ must have an analogue within the definition as a derivation in $D_p\mathcal{M}$, and *vice versa*. In particular, any tangent vector $v \in T_p\mathcal{M}$

can be expanded in components by thinking of it as a derivation and then using Eq. (2.3.30) to give

$$v = \sum_{\mu=1}^{m} v^\mu \left(\frac{\partial}{\partial x^\mu}\right)_p \tag{2.3.38}$$

with

$$v^\mu := v(x^\mu) = \frac{d}{dt} x^\mu \circ \sigma(t)\Big|_{t=0} \tag{2.3.39}$$

where $[\sigma] = v$.

3. Similarly, within the algebraic language of derivations, the push-forward map $h_*$—defined for equivalence classes of curves by Eq. (2.3.8)—acquires a new form. Specifically, if $h : \mathcal{M} \to \mathcal{N}$, then $h_* : D_p\mathcal{M} \to D_{h(p)}\mathcal{N}$ is defined on any $v \in D_p\mathcal{M}$ as

$$(h_* v)(f) := v(f \circ h) \tag{2.3.40}$$

for all $f \in C^\infty(\mathcal{N})$. Note that, in this form, the map $h_*$ is manifestly linear—a good example of the general maxim that algebraic properties of tangent spaces are best seen in the algebraic approach.

4. A useful example of an application of these results is to find a concrete expression for the local representative of the push-forward map $h_* : T_p\mathcal{M} \simeq D_p\mathcal{M} \to T_{h(p)}\mathcal{N} \simeq D_{h(p)}\mathcal{N}$ induced by a smooth map $h : \mathcal{M} \to \mathcal{N}$.

Specifically, suppose the manifolds $\mathcal{M}$ and $\mathcal{N}$ have dimension $m$ and $n$ respectively, and let $\{x^1, x^2, \ldots, x^m\}$ and $\{y^1, y^2, \ldots, y^n\}$ be local coordinate systems around the points $p \in \mathcal{M}$ and $h(p) \in \mathcal{N}$ respectively. According to Eq. (2.3.38), if $v \in T_p\mathcal{M}$, we can write $v = \sum_{\mu=1}^m v^\mu \left(\frac{\partial}{\partial x^\mu}\right)_p$ where $v^\mu = v(x^\mu)$, $\mu = 1, 2, \ldots m$; and then, since $h_* : D_p\mathcal{M} \to\simeq D_{h(p)}\mathcal{N}$ is linear, we have

$$h_* v = \sum_{\mu=1}^{m} v^\mu \, h_* \left(\frac{\partial}{\partial x^\mu}\right)_p . \tag{2.3.41}$$

On the other hand, the local coordinate system around $h(p) \in \mathcal{N}$ can be used to expand the vector $h_* \left(\frac{\partial}{\partial x^\mu}\right)_p \in T_{h(p)}\mathcal{N}$ as

$$h_* \left(\frac{\partial}{\partial x^\mu}\right)_p = \sum_{\nu=1}^{n} \left(h_* \left(\frac{\partial}{\partial x^\mu}\right)_p\right)^\nu \left(\frac{\partial}{\partial y^\nu}\right)_{h(p)} . \tag{2.3.42}$$

However, according to the expansion result Eq. (2.3.30) applied to the tangent space $D_{h(p)}\mathcal{N}$, we have

$$\left(h_*\left(\frac{\partial}{\partial x^\mu}\right)_p\right)^\nu = \left(h_*\left(\frac{\partial}{\partial x^\mu}\right)_p\right)(y^\nu)$$

$$= \left(\frac{\partial}{\partial x^\mu}\right)_p (y^\nu \circ h) \tag{2.3.43}$$

$$= \left(\frac{\partial}{\partial x^\mu}\right)_p h^\nu \tag{2.3.44}$$

where Eq. (2.3.43) follows from Eq. (2.3.40), and where

$$h^\nu := y^\nu \circ h \tag{2.3.45}$$

$h^\nu = 1, 2 \ldots, n$ are real-valued functions (the *components* of the map $h$) defined on the coordinate neighbourhood of the point $p \in \mathcal{M}$. Note that the expression in Eq. (2.3.44)—which is called the *Jacobian matrix* of the function $h$ with respect to the given local coordinates on $\mathcal{M}$ and $\mathcal{N}$—is related via Eq. (2.3.19) to the standard partial derivative of the local representatives of these functions with respect to the coordinates $x^\mu$.

Finally, inserting Eq. (2.3.44) into Eq. (2.3.41), we get the following expression for the local representation of $h_*v \in D_{h(p)}\mathcal{N}$:

$$h_*v = \sum_{\mu=1}^{m}\sum_{\nu=1}^{n} v^\mu \left(\frac{\partial}{\partial x^\mu}\right)_p h^\nu \left(\frac{\partial}{\partial y^\nu}\right)_{h(p)}. \tag{2.3.46}$$

Note that, in this particular context, the precise expression $\left(\frac{\partial}{\partial x^\mu}\right)_p h^\nu$ is often written, rather heuristically, as $\partial h^\nu(p)/\partial x^\mu$, in terms of which Eq. (2.3.46) becomes

$$h_*v = \sum_{\mu=1}^{m}\sum_{\nu=1}^{n} v^\mu \frac{\partial h^\nu}{\partial x^\mu}(p) \left(\frac{\partial}{\partial y^\nu}\right)_{h(p)}. \tag{2.3.47}$$

5. Another example of the use of Eq. (2.3.40) is to a map $c : (-\epsilon, \epsilon) \to \mathcal{M}$, *i.e.*, a curve in $\mathcal{M}$. Then, for all $f \in C^\infty(\mathcal{M})$, Eq.

(2.3.40) implies

$$\left(c_*\left(\frac{d}{dt}\right)_{t=0}\right)f = \left(\frac{d}{dt}\right)_{t=0}(f\circ c) = \frac{d}{dt}(f(c(t)))\Big|_{t=0} = [c](f) \tag{2.3.48}$$

so that

$$c_*\left(\frac{d}{dt}\right)_0 = [c] \in T_p\mathcal{M} \tag{2.3.49}$$

where $\left(\frac{d}{dt}\right)_0$ denotes the derivation on $C^\infty(\mathbb{R})$ defined by

$$\left(\frac{d}{dt}\right)_0 f := \frac{df(t)}{dt}\Big|_{t=0} \tag{2.3.50}$$

for all $f \in C^\infty(\mathbb{R})$.

6. It can now be seen that the tangent bundle $T\mathcal{M}$ has a natural structure of a $2m$-dimensional differentiable manifold. Specifically, the $2m$ coordinates of a vector $v \in T_p\mathcal{M}$ are defined to be (i) the $m$ real numbers $\{x^1(p), x^2(p), \ldots, x^m(p)\}$ that fix the point $p$ in a local coordinate system on $\mathcal{M}$; and (ii) the $m$ real numbers $\{v(x^1), v(x^2), \ldots, v(x^m)\}$ that are the components of the derivation $v$ with respect to this coordinate system. Put more abstractly, if $(U, \phi)$ is a local coordinate chart around $p \in \mathcal{M}$ then define the bijection

$$\begin{aligned} T\mathcal{M}_{|U} &\rightarrow \phi(U)\times\mathbb{R}^m \subset \mathbb{R}^m\times\mathbb{R}^m \\ [c] &\mapsto (\phi(p), \phi_*[c]) \text{ where } [c]\in T_p\mathcal{M}, \end{aligned} \tag{2.3.51}$$

where $T\mathcal{M}_{|U}$ denotes the 'restriction' of $T\mathcal{M}$ to the subset $U \subset \mathcal{M}$. Note that in writing Eq.. (2.3.51), the Euclidean space $\mathbb{R}^m$ has been identified with any of its tangent spaces $T_{\vec{x}}\mathbb{R}^m$ by associating with $\vec{w} \in \mathbb{R}^m$ the derivation $\chi(\vec{w})$ in $D_{\vec{x}}\mathbb{R}^m$ defined by

$$\chi(\vec{w})(f) := \frac{d}{dt}f(\vec{x}+t\vec{w})\Big|_{t=0}, \tag{2.3.52}$$

*i.e.*, $\vec{w} \in \mathbb{R}^m$ generates the curve $t \mapsto \vec{x} + t\vec{w}$ in $\mathbb{R}^m$. This identification is discussed in more detail in the following subsection. □

### 2.3.6 The tangent space $T_vV$ of a vector space $V$

It is intuitively obvious that the tangent space at any point in a vector space $V$ considered as a manifold, is isomorphic to $V$ itself in some natural way.

To show this explicitly, let us think of $T_vV$ as the set of derivations $D_vV$ at $v \in V$ of the ring of smooth functions $C^\infty(V)$. Then we can define a map $\chi : V \to D_vV$ by

$$\chi(w)f := \frac{d}{dt} f(v + tw)|_{t=0} \tag{2.3.53}$$

for all $f \in C^\infty(V)$. We must now show that $\chi$ is linear, one-to-one, and onto.

To show linearity, we note that

$$\begin{aligned} \chi(w_1 + w_2)f &= \frac{d}{dt} f(v + tw_1 + tw_2)|_{t=0} \\ &= \frac{d}{dt} f(v + tw_1)|_{t=0} + \frac{d}{dt} f(v + tw_2)|_{t=0} \\ &= \chi(w_1)f + \chi(w_2)f. \end{aligned} \tag{2.3.54}$$

Also, for all $r \in \mathbb{R}$,

$$\begin{aligned} \chi(rw)f &= \frac{d}{dt} f(v + trw)|_{t=0} = r\frac{d}{dt} f(v + tw)|_{t=0} \\ &= r\chi(w)f, \end{aligned} \tag{2.3.55}$$

and hence $\chi : V \to D_vV$ is linear.

To show that $\chi$ is one-to-one, suppose that $w_1, w_2 \in V$ are such that $\chi(w_1) = \chi(w_2)$; *i.e.*, $\chi(w_1 - w_2)$=0. Then, for all $f \in C^\infty(V)$, we have

$$\frac{d}{dt} f(v + t(w_1 - w_2))|_{t=0} = 0. \tag{2.3.56}$$

In particular, this is true for the function on $V$ given by any linear map $\ell : V \to \mathbb{R}$, and then

$$0 = \frac{d}{dt} \left(\ell(v) + t\ell(w_1 - w_2)\right)|_{t=0} = \ell(w_1 - w_2). \tag{2.3.57}$$

However, the linear maps on a finite-vector space 'separate' the points in the space in the sense that if $\ell(w) = 0$ for all such linear maps $\ell$, then $w = 0$. Thus Eq. (2.3.57) implies that $w_1 = w_2$; and hence $\chi$ is one-to-one.

Finally, to show that $\chi$ is onto, it suffices to note that any equivalence class of curves through $v$ clearly contains one of the form $t \mapsto v + tw$ for some $w \in V$. Then we invoke the isomorphism of the space of equivalence classes of curves with the space of derivations at the point $v \in V$.

### 2.3.7 A simple example of the push-forward operation

It may be helpful to illustrate the ideas of the push-forward operation with the aid of a simple example of a map between two vector spaces considered as differentiable manifolds.

**Problem**

The map $h : \mathbb{R}^2 \to \mathbb{R}^2$ is defined by

$$h(a, b) := (a^2 - 2b, 4a^3b^2). \tag{2.3.58}$$

Compute the push-forward of the derivation $h_*(4x^1\frac{\partial}{\partial x^1} + 3x^2\frac{\partial}{\partial x^2})_{(1,2)}$ at the point $(1, 2) \in \mathbb{R}^2$, where $\{x^1, x^2\}$ are the natural coordinates on $\mathbb{R}^2$.

**Answer**

If we use the natural global coordinates $\{x^1, x^2\}$ on the domain manifold $\mathbb{R}^2$ of the map $h : \mathbb{R}^2 \to \mathbb{R}^2$, the components of the vector field $v := 4x^1\frac{\partial}{\partial x^1} + 3x^2\frac{\partial}{\partial x^2}$ are $v^1 = 4x^1$ and $v^2 = 3x^2$, which have the values 4 and 6 respectively at the specific point $(1, 2)$.

To avoid confusion, it is advisable to use a different letter of the alphabet to denote the global coordinates on the target manifold $\mathbb{R}^2$ from that used $(x)$ on the domain manifold $\mathbb{R}^2$. To this end, we denote the natural global coordinates on the target manifold $\mathbb{R}^2$ by

$\{y^1, y^2\}$, and then the components of the map $h$ are

$$\begin{aligned} h^1(a,b) &:= y^1 \circ h(a,b) = a^2 - 2b \\ h^2(a,b) &:= y^2 \circ h(a,b) = 4a^3b^2. \end{aligned} \tag{2.3.59}$$

Indeed, this expression is what is really meant by the notation used in the definition Eq. (2.3.58) of the function $h$. In particular, the point with $x$-coordinates $(1,2)$ is mapped by $h$ to the point with $y$-coordinates $(-3,16)$.

The Jacobian matrix of the transformation is

$$\begin{array}{ll} (\partial h^1/\partial x^1)(a,b) = 2a, & (\partial h^1/\partial x^2)(a,b) = -2 \\ (\partial h^2/\partial x^1)(a,b) = 12a^2b^2, & (\partial h^2/\partial x^2)(a,b) = 8a^3b \end{array} \tag{2.3.60}$$

at any point $(a,b)$ in the domain manifold $\mathbb{R}^2$. Thus, according to Eq. (2.3.47), the components of $h_*v$ are

$$(h_*v)^1 = v^1(\partial h^1/\partial x^1)(1,2) + v^2(\partial h^1/\partial x^2)(1,2) = 4\times 2 + 6\times(-2) = -4 \tag{2.3.61}$$

and

$$(h_*v)^2 = v^1(\partial h^2/\partial x^1)(1,2) + v^2(\partial h^2/\partial x^2)(1,2) = 4\times 48 + 6\times 16 = 288, \tag{2.3.62}$$

so that

$$h_*v = -4\left(\frac{\partial}{\partial y^1}\right)_{(-3,16)} + 288\left(\frac{\partial}{\partial y^2}\right)_{(-3,16)}. \tag{2.3.63}$$

### 2.3.8 The tangent space of a product manifold

We come now to some technical results concerning the tangent space structure of the product $\mathcal{M} \times \mathcal{N}$ of two manifolds $\mathcal{M}$ and $\mathcal{N}$. These will be of importance in our discussions later of Lie groups and fibre bundles.

If one takes the torus $S^1 \times S^1$ as an example, it is intuitively clear that a tangent vector at a point $(p,q)$ in a product $\mathcal{M} \times \mathcal{N}$ can be 'decomposed' into a component that lies 'along' $\mathcal{M}$, and another that lies along $\mathcal{N}$. The formal expression of this intuition is contained in the following theorem.

**Theorem 2.3** *On the product manifold $\mathcal{M} \times \mathcal{N}$ there is a natural isomorphism*[10]

$$\begin{aligned} T_{(p,q)}(\mathcal{M} \times \mathcal{N}) &\simeq T_p\mathcal{M} \oplus T_q\mathcal{N} \\ v &\mapsto (\mathrm{pr}_{1*}(v), \mathrm{pr}_{2*}(v)) \end{aligned} \qquad (2.3.64)$$

*where*

$$\begin{aligned} \mathrm{pr}_1 : \mathcal{M} \times \mathcal{N} &\to \mathcal{M}; \qquad (2.3.65) \\ (p,q) &\mapsto p \\ \mathrm{pr}_2 : \mathcal{M} \times \mathcal{N} &\to \mathcal{N} \qquad (2.3.66) \\ (p,q) &\mapsto q \end{aligned}$$

*are the usual projection maps.*

## Proof

Let $\sigma : (-\epsilon, \epsilon) \to \mathcal{M} \times \mathcal{N}$ be a curve such that $[\sigma] = v \in T_{(p,q)}\mathcal{M} \times \mathcal{N}$. Then there exist unique curves $\sigma_1$ and $\sigma_2$ in $\mathcal{M}$ and $\mathcal{N}$ respectively such that $\sigma(t) = (\sigma_1(t), \sigma_2(t))$: namely, $\sigma_1(t) := \mathrm{pr}_1 \circ \sigma(t)$ and $\sigma_2(t) := \mathrm{pr}_2 \circ \sigma(t)$.

Define a linear map $\chi : T_{(p,q)}\mathcal{M} \times \mathcal{N} \to T_p\mathcal{M} \oplus T_q\mathcal{N}$ by

$$\chi([\sigma]) := ([\sigma_1], [\sigma_2]). \qquad (2.3.67)$$

This is clearly one-to-one and surjective, and is hence the desired isomorphism. Furthermore,

$$[\sigma_1] = \sigma_{1*}\left(\frac{d}{dt}\right)_0 = (\mathrm{pr}_1 \circ \sigma)_*\left(\frac{d}{dt}\right)_0 = \mathrm{pr}_{1*}\sigma_*\left(\frac{d}{dt}\right)_0 = \mathrm{pr}_{1*}[\sigma] \qquad (2.3.68)$$

and so $[\sigma_1] = \mathrm{pr}_{1*}[\sigma]$, as claimed (where Eq. (2.3.49) has been used).

**QED**

---

[10]The direct sum $V \oplus W$ of two real vector spaces $V$ and $W$ is defined to be the set of all pairs $(v, w)$ with $v \in V$ and $w \in W$, and where the vector space operations are defined as (i) $(v_1, w_1) + (v_2, w_2) := (v_1 + v_2, w_1 + w_2)$; and (ii) $r(v, w) := (rv, rw)$ for $r \in \mathbb{R}$.

**Comments**

1. It is convenient to define the injections (for each $p \in \mathcal{M}$ and $q \in \mathcal{N}$):

$$\begin{aligned} i_q : \mathcal{M} &\to \mathcal{M} \times \mathcal{N}; \\ x &\mapsto (x, q) \end{aligned} \tag{2.3.69}$$

$$\begin{aligned} j_p : \mathcal{N} &\to \mathcal{M} \times \mathcal{N}; \\ y &\mapsto (p, y) \end{aligned} \tag{2.3.70}$$

which satisfy the relations

$$\mathrm{pr}_1 \circ i_q = \mathrm{id}_{\mathcal{M}} \quad ; \quad \mathrm{pr}_2 \circ i_q = q(\,) \tag{2.3.71}$$

$$\mathrm{pr}_1 \circ j_p = p(\,) \quad ; \quad \mathrm{pr}_2 \circ j_p = \mathrm{id}_{\mathcal{N}} \tag{2.3.72}$$

where $\mathrm{id}_{\mathcal{M}}$ (resp. $\mathrm{id}_{\mathcal{N}}$) is the identity map of $\mathcal{M}$ (resp. $\mathcal{N}$) into itself, and $q(\,)$ (resp. $p(\,)$) is the constant map on $\mathcal{N}$ (resp. $\mathcal{M}$) that takes every point into $q \in \mathcal{N}$ (resp. $p \in \mathcal{M}$).

Thus the inverse of the isomorphism $\chi$ in Eq. (2.3.67) is the map

$$\begin{aligned} T_p\mathcal{M} \oplus T_q\mathcal{N} &\to T_{(p,q)}\mathcal{M} \times \mathcal{N}. \\ (\alpha, \beta) &\mapsto i_{q*}\alpha + j_{p*}\beta \end{aligned} \tag{2.3.73}$$

2. Let $\mathcal{P}$ be another manifold and let $f : \mathcal{M} \times \mathcal{N} \to \mathcal{P}$. Then $f_* : T_{(p,q)}\mathcal{M} \times \mathcal{N} \to T_{f(p,q)}\mathcal{P}$ factorises as

$$T_{(p,q)}\mathcal{M} \times \mathcal{N} \xrightarrow{\chi} T_p\mathcal{M} \oplus T_q\mathcal{N} \xrightarrow{\tilde{f}} T_{f(p,q)}\mathcal{P} \tag{2.3.74}$$

where

$$\tilde{f}(\alpha, \beta) := (f \circ i_q)_*\alpha + (f \circ j_p)_*\beta \tag{2.3.75}$$

and

$$\begin{aligned} f \circ i_q : \mathcal{M} &\to \mathcal{P} \\ x &\mapsto f(x, q) \end{aligned} \tag{2.3.76}$$

$$\begin{aligned} f \circ j_p : \mathcal{N} &\to \mathcal{P}. \\ y &\mapsto f(p, y) \end{aligned} \tag{2.3.77}$$

3. The *diagonal* map associated with a manifold $\mathcal{M}$ is the map $\Delta : \mathcal{M} \to \mathcal{M} \times \mathcal{M}$ defined by $\Delta(p) := (p, p)$. In the chain of maps

$$T_p\mathcal{M} \xrightarrow{\Delta_*} T_{(p,p)}\mathcal{M} \times \mathcal{M} \xrightarrow{\chi} T_p\mathcal{M} \oplus T_p\mathcal{M} \tag{2.3.78}$$

we have

$$\chi \circ \Delta_*(\alpha) = (\alpha, \alpha) \tag{2.3.79}$$

for all $\alpha \in T_p\mathcal{M}$.

In particular, if $f : \mathcal{M} \times \mathcal{M} \to \mathcal{P}$ then $f \circ \Delta : \mathcal{M} \to \mathcal{P}$ with $f \circ \Delta(p) = f(p, p)$ and $(f \circ \Delta)_*(\alpha) = f_*(\Delta_*(\alpha)) = \tilde{f} \circ \chi \circ \Delta_*(\alpha) = \tilde{f}(\alpha, \alpha)$ for all $\alpha \in T_p\mathcal{M}$. Thus, on $T_p\mathcal{M}$,

$$(f \circ \Delta)_* = (f \circ i_p)_* + (f \circ j_p)_*. \tag{2.3.80}$$

# Chapter 3

# Vector Fields and $n$-Forms

## 3.1 Vector Fields

### 3.1.1 The main definition

So far we have dealt only with tangent vectors defined at a single point in the manifold. We come now to an important generalisation of this idea that involves a 'field' of tangent vectors in the sense that a tangent vector is assigned to *each* point of $\mathcal{M}$ (or, more generally, to each point in some open subset of $\mathcal{M}$). A critical requirement here is that the tangent vectors should vary 'smoothly' as we move around the manifold—a concept that is formalised in the following definition.

**Definition 3.1**

1. A *vector field* $X$ on a $C^\infty$-manifold $\mathcal{M}$ is a smooth assignment of a tangent vector $X_p \in T_p\mathcal{M}$ at each point $p \in \mathcal{M}$, where 'smooth' is defined to mean that, for all $f \in C^\infty(\mathcal{M})$, the function $Xf : \mathcal{M} \to \mathbb{R}$ defined by

$$\begin{aligned} \mathcal{M} &\to \mathbb{R} \\ p &\mapsto (Xf)(p) := X_p(f) \end{aligned} \tag{3.1.1}$$

   is infinitely differentiable.

2. A vector field on an open subset $U$ of $\mathcal{M}$ is defined in the same way except that the condition in Eq. (3.1.1) is required to be true for all $f \in C^\infty(U)$ and points $p$ in the subset $U \subset \mathcal{M}$.

## Comments

1. A vector field $X$ on $\mathcal{M}$ gives rise in an obvious way to a vector field, denoted $X_U$, on any open subset $U$ of $\mathcal{M}$. The vector field $X_U$ is said to be the *restriction* of $X$ to $U$.

2. A vector field on $\mathcal{M}$ can also be defined as a smooth 'cross-section' of the tangent bundle $T\mathcal{M}$:

$$\begin{array}{c} T\mathcal{M} \\ \downarrow^{\pi} \quad \uparrow^{X} \\ \mathcal{M} \end{array} \qquad \pi \circ X = \mathrm{id}_{\mathcal{M}}. \tag{3.1.2}$$

In this case, 'smooth' means that $X$ is a $C^\infty$-map from the manifold $\mathcal{M}$ to the $2m$-dimensional manifold $T\mathcal{M}$ (using the differential structure defined in relation to Eq. (2.3.51)). It is straightforward to show that this definition of a smooth assignment agrees with the one above.

3. A vector field defined via Eq. (3.1.1) gives rise to a map $X : C^\infty(\mathcal{M}) \to C^\infty(\mathcal{M})$ in which[1] the image $Xf$ of $f \in C^\infty(\mathcal{M})$ is defined by Eq. (3.1.1). The function $Xf$ is often known as the *Lie derivative* of the function $f$ along the vector field $X$, and is then usually denoted $L_X f$, or sometimes $\pounds_X f$.

4. The map $f \mapsto Xf$ has the following properties with respect to the ring structure on $C^\infty(\mathcal{M})$:

$$\text{(i) } X(f+g) = Xf + Xg \text{ for all } f, g \in C^\infty(\mathcal{M}) \tag{3.1.3}$$

$$\text{(ii) } X(rf) = rXf \text{ for all } f \in C^\infty(\mathcal{M}),\ r \in \mathbb{R} \tag{3.1.4}$$

$$\text{(iii) } X(fg) = fXg + gXf \text{ for all } f, g \in C^\infty(\mathcal{M}) \tag{3.1.5}$$

[1]The map $X : C^\infty(\mathcal{M}) \to C^\infty(\mathcal{M})$ is *linear* and—as with all linear maps from a vector space to itself (*i.e.*, linear operators)—it is conventional to write the image $X(f)$ of $f \in C^\infty(\mathcal{M})$ simply as $Xf$, *i.e.*, the brackets are usually left out for linear maps unless it is necessary to add them for clarity.

which should be contrasted with the equations Eqs. (2.3.14–2.3.16) for a derivation in $D_p\mathcal{M}$. Note that (i) Eqs. (3.1.3–3.1.4) show that $X$ is a *linear* map from the vector space $C^\infty(\mathcal{M})$ into itself; and (ii) Eq. (3.1.5) shows that $X$ is a *derivation* of the ring $C^\infty(\mathcal{M})$ in the usual sense of the word (in this respect, the use of the term 'derivation' in relation to Eq. (2.3.16) is potentially confusing, but it is conventional).

5. This procedure can be reversed in the sense that it is possible to *define* a vector field $X$ as a derivation of the ring of functions $C^\infty(\mathcal{M})$, *i.e.*, a map from $C^\infty(\mathcal{M})$ to $C^\infty(\mathcal{M})$ that satisfies the three equations Eqs. (3.1.3–3.1.5). Indeed, to each point $p$ in $\mathcal{M}$ such a derivation assigns a linear map $X_p : C^\infty(\mathcal{M}) \to \mathbb{R}$ defined by

$$X_p(f) := (Xf)(p) \tag{3.1.6}$$

for each $f$ in $C^\infty(\mathcal{M})$, and it is easy to see that $X_p$ satisfies the three conditions Eqs. (2.3.14–2.3.16) for a derivation at a point. But then the map $p \mapsto X_p$ assigns a field of tangent vectors in the sense of Definition 3.1, and it is clear that this assignment is smooth in the sense of Eq. (3.1.1).

This alternative approach—in which we go from vector field to derivation at a point $p \in \mathcal{M}$, rather than the converse—is useful in many situations, and we shall employ it where appropriate.

6. The set of all vector fields on a manifold $\mathcal{M}$ is denoted VFld$(\mathcal{M})$ and carries a natural structure of a real vector space. More precisely, if $X, Y$ are vector fields on $\mathcal{M}$ and $a, b \in \mathbb{R}$ then $aX + bY$ is defined as

$$(aX + bY)f := aXf + bYf \tag{3.1.7}$$

for all $f \in C^\infty(\mathcal{M})$.

It is important to note that this definition can be extended to give VFld$(\mathcal{M})$ the structure of a *module* over the ring $C^\infty(\mathcal{M})$. Specifically, if $g, h \in C^\infty(\mathcal{M})$ and if $X$ and $Y$ are vector fields on $\mathcal{M}$, a new vector field $gX + hY$ can be defined by

$$(gX + hY)_p f := g(p)X_p f + h(p)Y_p f \tag{3.1.8}$$

for all $p \in \mathcal{M}$ and $f \in C^\infty(\mathcal{M})$.

7. Let $(U, \phi)$ be a coordinate chart on the manifold $\mathcal{M}$. Then at each point $p$ in the open set $U$ we can use the expansion Eq. (2.3.30) for the derivation $X_p$ associated with a vector field $X$ defined on $\mathcal{M}$. Thus, for all $p \in U$,

$$\begin{aligned}(Xf)(p) = X_p f &= \sum_{\mu=1}^{m} X_p x^\mu \left(\frac{\partial}{\partial x^\mu}\right)_p f \\ &= \sum_{\mu=1}^{m} (Xx^\mu)(p) \left(\frac{\partial}{\partial x^\mu}\right)_p f \qquad (3.1.9)\end{aligned}$$

which we write as

$$X_U = \sum_{\mu=1}^{m} X_U x^\mu \frac{\partial}{\partial x^\mu} \qquad (3.1.10)$$

in which the module structure on VFld$(U)$ is used to expand the vector field $X_U$ in terms of the local coordinate vector fields $\frac{\partial}{\partial x^\mu}$, $\mu = 1, 2, \ldots, m$ where the coefficient functions $Xx^\mu$ are defined on the open subset $U$. It should be noted that the subscript $U$ is often dropped, so that Eq. (3.1.10) becomes simply

$$X = \sum_{\mu=1}^{m} Xx^\mu \frac{\partial}{\partial x^\mu} \qquad (3.1.11)$$

where there is an implicit understanding that this equation is meaningful only on the open set on which the coordinates $\{x^\mu\}$ are defined.

The expression Eq. (3.1.10) (interpreted as Eq. (3.1.9)) shows the precise sense in which a vector field can be thought of as a first-order differential operator on the functions on a manifold.

8. The functions $\{Xx^\mu, \mu = 1, 2 \ldots, m\}$ on $U$ are known as the *components* of the vector field $X$ with respect to the coordinate system associated with the chart $(U, \phi)$. They are usually denoted $X^\mu$.

Of course, the actual form of these component functions depends on the coordinate chart in use. Specifically, let $(U', \phi')$ be another coordinate chart with the property that $U \bigcap U' \neq \emptyset$ so that the components $Xx'^\nu$ and $Xx^\mu$ can be meaningfully compared on the open set $U \bigcap U'$. Then using Eq. (3.1.11) with respect to both systems of coordinates gives, on the open subset $U \bigcap U' \subset \mathcal{M}$,

$$X = \sum_{\mu=1}^{m} X x^\mu \frac{\partial}{\partial x^\mu} = \sum_{\nu=1}^{m} X x'^\nu \frac{\partial}{\partial x'^\nu} \tag{3.1.12}$$

and Eq. (3.1.11) can be applied again—this time to the vector field $\partial/\partial x^\mu$ expanded with respect to the coordinates $x'^\nu$—to get

$$\frac{\partial}{\partial x^\mu} = \sum_{\nu=1}^{m} \frac{\partial}{\partial x^\mu}(x'^\nu)\frac{\partial}{\partial x'^\nu}. \tag{3.1.13}$$

The components of $X$ in the two coordinate systems are $X^\mu := X x^\mu$ and $X^{\nu'} := X x'^\nu$ respectively. Then, substituting Eq. (3.1.13) into Eq. (3.1.12) and equating terms gives

$$\boxed{X^{\nu'} = \sum_{\mu=1}^{m} X^\mu \frac{\partial x'^\nu}{\partial x^\mu}} \tag{3.1.14}$$

which is a familiar expression in elementary, coordinate-based approaches to differential geometry. For this reason, Eq. (3.1.11) plays a key role in understanding the precise connections between older approaches to differential geometry and the more modern, coordinate-free formalism that is being developed here. In this context it is important to appreciate that the precise meaning of the expression $\partial x'^\nu/\partial x^\mu$ in Eq. (3.1.14) is the function on the open set $U \bigcap U'$ defined by

$$\frac{\partial x'^\nu}{\partial x^\mu}(p) := \left(\frac{\partial}{\partial x^\mu}\right)_p x'^\nu = \frac{\partial}{\partial u^\mu} x'^\nu \circ \phi^{-1}\bigg|_{\phi(p)}. \tag{3.1.15}$$

9. One of the important properties of vector fields is their ability to encode certain global, topological properties of the manifold. A famous example is the theorem showing that a necessary condition for the existence of a nowhere-vanishing vector field $X$ (*i.e.*, $X_p \neq 0$ for all $p \in \mathcal{M}$) is that the Euler number of the manifold $\mathcal{M}$ is zero (for example, see Bott & Tu (1982)). This is nicely illustrated by the 2-sphere $S^2$: this has Euler number 2, and all attempts to construct a nowhere-vanishing vector field necessarily fail (this is known colloquially as the 'hairy billiard-ball problem').

On the other hand, the 2-torus $S^1 \times S^1$ has Euler number 0, and it is easy to see that nowhere-vanishing vector fields exist on this space. □

### 3.1.2 The vector field commutator

We come now to the rather important question of whether or not two vector fields $X$ and $Y$ can be 'multiplied' together to get a third field. Since $X$ and $Y$ can be viewed as linear maps of $C^\infty(\mathcal{M})$ into itself, the composite map $X \circ Y : C^\infty(\mathcal{M}) \to C^\infty(\mathcal{M})$ can be defined as $X \circ Y(f) := X(Yf)$ and this is certainly linear, *i.e.*, it satisfies Eqs. (3.1.3–3.1.4). However, it is *not* a vector field since it fails to satisfy the crucial derivation property in Eq. (3.1.5). Indeed,

$$\begin{aligned} X \circ Y(fg) &= X(Y(fg)) = X(gYf + fYg) \\ &= XgYf + gX(Yf) + XfYg + fX(Yg) \end{aligned} \quad (3.1.16)$$

which does not equal $gX(Yf) + fX(Yg)$ as it would have to if it was to be a derivation. However, we also have

$$Y \circ X(fg) = YgXf + gY(Xf) + YfXg + fY(Xg) \quad (3.1.17)$$

and subtracting Eq. (3.1.17) from Eq. (3.1.16) gives

$$(X \circ Y - Y \circ X)(fg) = g(X \circ Y - Y \circ X)f + f(X \circ Y - Y \circ X)g \quad (3.1.18)$$

which illustrates the vital point that $X \circ Y - Y \circ X$ *is* a vector field even though the individual composites $X \circ Y$ and $Y \circ X$ are not.

This new vector field $X \circ Y - Y \circ X$ is called the *commutator* of the vector fields $X$ and $Y$ and is written as $[X, Y]$.

#### Comments

1. In terms of the local representation Eq. (3.1.11) of a vector field as a first-order partial-differential operator it is clear that the simple composition $X \circ Y$ cannot be a vector field since it includes second-order differential operators as well as those of first order. When $Y \circ X$ is subtracted from $X \circ Y$ these second-order terms cancel and only

the first-order differential operators are left. The components of the commutator $[X, Y]$ are readily computed [Exercise!] to be

$$[X, Y]^\mu = \sum_{\nu=1}^{m} (X^\nu Y^\mu{}_{,\nu} - Y^\nu X^\mu{}_{,\nu}) \tag{3.1.19}$$

where, as is conventional, the expression $Y^\mu{}_{,\nu}$ denotes the derivative of the function $Y^\mu$ with respect to the coordinate $x^\nu$ in the sense of Eq. (3.1.15).

2. The algebraic pairing of vector fields $X, Y$ to give a new field $[X, Y]$ is clearly antisymmetric in the sense that

$$[X, Y] = -[Y, X]. \tag{3.1.20}$$

3. If $X$,$Y$ and $Z$ are any three vector fields on $\mathcal{M}$ then their commutators satisfy the so-called *Jacobi identity*:

$$[X, [Y, Z]] + [Z, [X, Y]] + [Y, [Z, X]] = 0. \tag{3.1.21}$$

4. The vector field commutator is related in an important way to the idea of a Lie algebra. In general, a real *Lie algebra* is defined to be a real vector space $\mathcal{L}$ with a bilinear map $\mathcal{L} \times \mathcal{L} \to \mathcal{L}$, denoted $(A, B) \mapsto [AB]$, which satisfies the following conditions:

*Antisymmetry*: For all $A, B \in \mathcal{L}$,

$$[AB] = -[AB] \tag{3.1.22}$$

*Jacobi identity*: For all $A, B, C \in \mathcal{L}$,

$$[A[BC]] + [B[CA]] + [C[AB]] = 0. \tag{3.1.23}$$

If the operation $(A, B) \mapsto [AB]$ is thought of a type of 'multiplication', the condition Eq. (3.1.23) shows that it is not associative. Note that the word 'identity' is somewhat misplaced here: Eq. (3.1.23) is not an identity in the strict sense; it is just one of the conditions that the multiplication operation has to satisfy in order to be a Lie algebra.

An important example of a Lie algebra is the set $M(n, \mathbb{R})$ of all $n \times n$ real matrices, with $[AB] := AB - BA$. In this case, Eq. (3.1.23) *is* a genuine 'identity'; *i.e.*, the equation $[A, [B, C]] + [B, [C, A]] + [C, [A, B]] = 0$ is automatically satisfied for all $A, B, C \in M(n, \mathbb{R})$ by virtue of the rule for matrix multiplication.

The two results in Eqs. (3.1.20–3.1.21) have the important implication that VFld($\mathcal{M}$) has the structure of a real Lie algebra, and—as in the matrix case—the equation Eq. (3.1.21) is a genuine identity. This algebra is infinite-dimensional and—with due care—can be regarded as the Lie algebra of the infinite-dimensional Lie group Diff($\mathcal{M}$) of diffeomorphisms of $\mathcal{M}$.

This result is related to the—somewhat heuristic—assertion that a vector field is a 'generator of infinitesimal transformations' on the manifold. We shall return later to this idea and also to the important question of when the infinite-dimensional Lie algebra VFld($\mathcal{M}$) possesses a non-trivial finite-dimensional subalgebra. □

### 3.1.3 *h*-related vector fields

In Section 2.3.5 we defined the push-forward $h_* : T_p\mathcal{M} \to T_{h(p)}\mathcal{N}$ induced by a map $h : \mathcal{M} \to \mathcal{N}$. This was a linear map between the vector spaces $T_p\mathcal{M}$ and $T_{h(p)}\mathcal{N}$, and the question arises of whether it is similarly possible to define an induced map from the vector fields on $\mathcal{M}$ to those on $\mathcal{N}$. Given a vector field $X$ on $\mathcal{M}$ and a smooth map $h$ from $\mathcal{M}$ to $\mathcal{N}$, a natural choice for an induced vector field $h_*X$ on $\mathcal{N}$ might appear to be:

$$(h_*X)_{h(p)} := h_*(X_p) \tag{3.1.24}$$

but, as can be seen with the aid of the following diagram (Figure 3.1), this may fail to be well-defined for two reasons:

(i) If there are points $p_1$ and $p_2$ such that $h(p_1) = h(p_2)$ (*i.e.*, if $h$ is not one-to-one) then the definition in Eq. (3.1.24) will be ambiguous if $h_*(X_{p_1}) \neq h_*(X_{p_2})$.

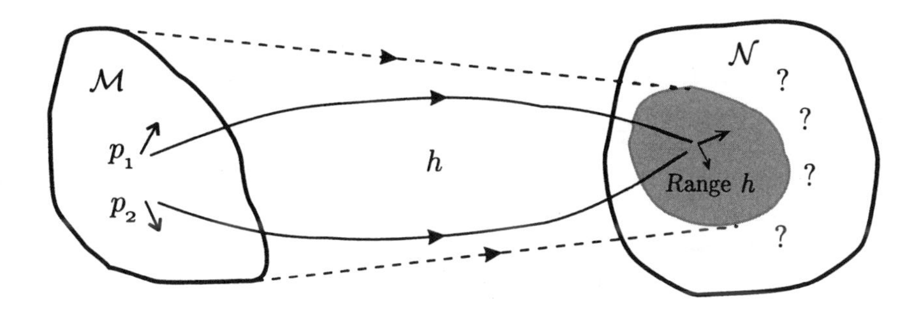


Figure 3.1: The problem of pushing-forward a vector field.

(ii) If $h$ is not surjective then Eq. (3.1.24) does not specify the induced field outside the range of $h$.

Note that if $h$ is a *diffeomorphism* from $\mathcal{M}$ to $\mathcal{N}$ then neither of these objections apply and an induced vector field $h_*X$ *can* be defined via Eq. (3.1.24). However, it is possible that in certain cases the idea will work, even if $h$ is not a diffeomorphism, and this motivates the following definition:

**Definition 3.2**

Let $h : \mathcal{M} \to \mathcal{N}$ and let $X$ and $Y$ be vector fields on $\mathcal{M}$ and $\mathcal{N}$ respectively. Then $X$ and $Y$ are said to be *h-related* if, at all points $p$ in $\mathcal{M}$,

$$h_*(X_p) = Y_{h(p)} \tag{3.1.25}$$

and we then write $Y = h_*X$.

**Comments**

1. If $X$ and $Y$ are $h$-related, and if $f \in C^\infty(\mathcal{N})$, then

$$(h_*X_p)f = Y_{h(p)}f \tag{3.1.26}$$

*i.e.*,

$$[X(f \circ h)](p) = X_p(f \circ h) = (Yf)(h(p)) \tag{3.1.27}$$

and so

$$X(f \circ h) = (Yf) \circ h, \tag{3.1.28}$$

which is a useful way in which to express the property of $h$-relatedness (see Figure 3.2).

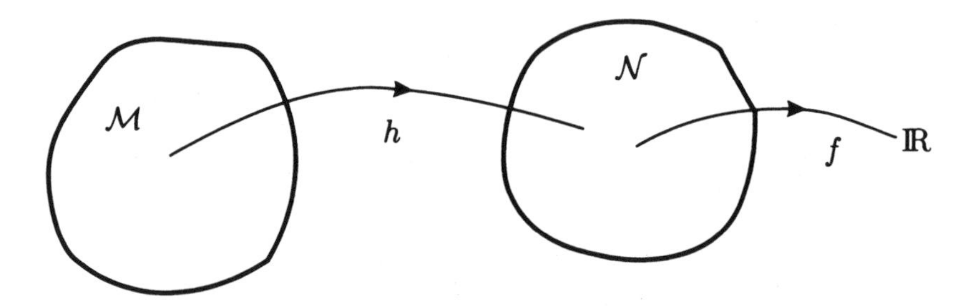


Figure 3.2: The functions involved in $h$-relatedness.

2. If $h_* X_1 = Y_1$ and $h_* X_2 = Y_2$ then, for all $f \in C^\infty(\mathcal{M})$, Eq. (3.1.27) can be used to write

$$\begin{aligned} ([Y_1, Y_2]f) \circ h &= (Y_1(Y_2 f)) \circ h - (Y_2(Y_1 f)) \circ h \\ &= X_1((Y_2 f) \circ h) - X_2((Y_1 f) \circ h) \\ &= X_1(X_2(f \circ h)) - X_2(X_1(f \circ h)) \\ &= [X_1, X_2](f \circ h), \end{aligned} \tag{3.1.29}$$

which means that

$$h_*([X_1, X_2]_p) = [h_* X_1, h_* X_2]_{h(p)} \tag{3.1.30}$$

for all $p \in \mathcal{M}$. Thus $[X_1, X_2]$ is $h$-related to $[Y_1, Y_2]$ with

$$h_*[X_1, X_2] = [h_* X_1, h_* X_2]. \tag{3.1.31}$$

As we shall see later, this relation plays an important role in the development of the theory of the Lie algebra associated with a Lie group.

# 3.2 Integral Curves and Flows

## 3.2.1 Complete vector fields

In this section we shall discuss the precise sense in which a vector field can be regarded as a generator of an 'infinitesimal' diffeomorphism of a manifold. Expressions such as

$$\delta x^{\mu} = \epsilon X^{\mu}(x) \tag{3.2.1}$$

appear frequently in the physics-oriented literature but the deceptive simplicity of this equation covers up a fair amount of subtlety.

We start by recalling that tangent vectors can be regarded either as equivalence classes of curves or as derivations defined at a point in the manifold. It seems intuitively clear that if we have a smooth, non-intersecting family of curves on a manifold then the tangent vectors at each point can be taken together to form a vector field, as shown in Figure 3.3.

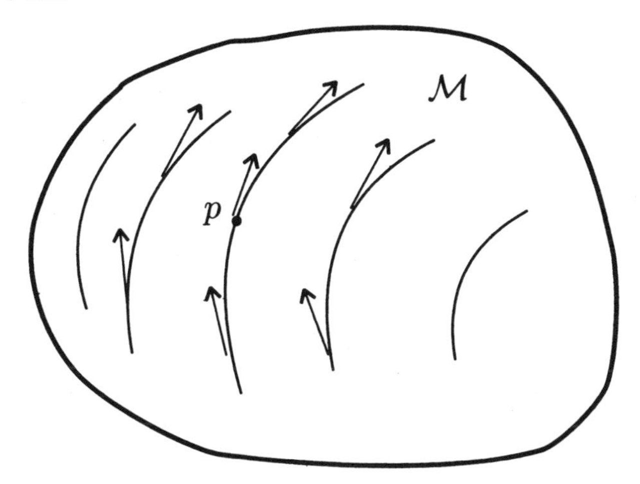


Figure 3.3: A field of tangent vectors.

More interesting however is the converse question. Namely, given a vector field $X$ on $\mathcal{M}$ is it possible to 'fill in' $\mathcal{M}$ with a family of curves in such a way that the tangent vector to the curve that passes through any particular point $p \in \mathcal{M}$ is just the vector field evaluated at that point, *i.e.*, the derivation $X_p$?

From a physical perspective, this idea is extremely important. For example, in the general canonical theory of classical mechanics, the state space of a system is represented by a certain (even-dimensional) manifold $\mathcal{S}$, and physical quantities are represented by real-valued differentiable functions on $\mathcal{S}$. The manifold $\mathcal{S}$ is equipped with a structure that associates to each such function $f : \mathcal{S} \to \mathbb{R}$ a vector field $X_f$ on $\mathcal{S}$, and the family of curves that 'fit' $X_f$ play an important role in the theory. In particular, the curves associated with the vector field $X_H$, where $H : \mathcal{S} \to \mathbb{R}$ is the *energy* function, are just the dynamical trajectories of the system.

Returning to the case of a general manifold $\mathcal{M}$, we consider first the simpler case of finding a single curve that (i) passes through a specified point $p \in \mathcal{M}$; and (ii) is such that the tangent vector at every point along this curve agrees with the vector field evaluated at that point. This idea is contained in the following definition.

**Definition 3.3**

Let $X$ be a vector field on a manifold $\mathcal{M}$ and let $p$ be a point in $\mathcal{M}$. Then an *integral curve* of $X$ passing through the point $p$ is a curve $t \mapsto \sigma(t)$ in $\mathcal{M}$ such that

$$\sigma(0) = p \tag{3.2.2}$$

and

$$\sigma_* \left(\frac{d}{dt}\right)_t = X_{\sigma(t)} \tag{3.2.3}$$

for all $t$ in some open interval $(-\epsilon, \epsilon)$ of $\mathbb{R}$.

**Comments**

1. In local coordinates, $X$ can be written as (see Eq. (3.1.11))

$$X = \sum_{\mu=1}^{m} X x^\mu \frac{\partial}{\partial x^\mu} \tag{3.2.4}$$

and, in the present case, Eq. (3.2.3) implies that

$$(Xx^\mu)(\sigma(t)) = X_{\sigma(t)}(x^\mu) = \sigma_* \left(\frac{d}{dt}\right)_t x^\mu = \frac{d}{dt} x^\mu \circ \sigma(t) \tag{3.2.5}$$

and hence the components $X^\mu$ of $X$ determine the integral curve $t \mapsto \sigma(t)$ according to the ordinary differential equation

$$X^\mu(\sigma(t)) = \frac{d}{dt} x^\mu(\sigma(t)) \quad \mu = 1, 2, \ldots, m \tag{3.2.6}$$

subject to the boundary condition

$$x^\mu(\sigma(0)) = x^\mu(p) \quad \mu = 1, 2, \ldots, m. \tag{3.2.7}$$

2. The discussion of the existence and uniqueness of solutions to the system of first-order differential equations in Eq. (3.2.6) with initial data Eq. (3.2.7) can be tackled by any of the familiar analytical methods. One approach is to employ the equivalent integral equation

$$x^\mu(\sigma(t)) = x^\mu(p) + \int_0^t X^\mu(\sigma(s))\, ds \tag{3.2.8}$$

which may be solved uniquely for some range of values for $t$ using, for example, the classical Picard iterative method or one of the more modern approaches based on fixed-point theorems. □

By some such means we can demonstrate the existence of an integral curve that passes through the specified point $p$ and is defined for $t$ lying in some open interval $(-\epsilon, \epsilon)$ of $\mathbb{R}$. However, it is a different matter when it comes to the existence of solutions that are defined for all values of the parameter $t$. Solutions in overlapping open sets can be patched together to some extent but this does not guarantee the existence for all values of $t \in \mathbb{R}$. For example, if $\mathcal{M}$ is non-compact a locally-defined integral curve may run off the 'edge' of $\mathcal{M}$ after only a finite range of $t$ values. This motivates the important definition:

**Definition 3.4**

> A vector field $X$ on a manifold $\mathcal{M}$ is *complete* if, at every point $p$ in $\mathcal{M}$, the integral curve that passes through $p$ can be extended to an integral curve for $X$ that is defined for all $t \in \mathbb{R}$.

**Comments**

1. It can be shown that if the manifold $\mathcal{M}$ is compact then any vector field is complete (Abraham & Marsden 1980). Since it is complete vector fields that can most readily be identified with the Lie algebra of the diffeomorphism group it follows that the study of Diff($\mathcal{M}$) is considerably easier when $\mathcal{M}$ is compact than when it is not.

2. When using differential geometry in theoretical physics, a situation that often arises is to have a vector field $X$ on a manifold $\mathcal{M}$ that is equipped with a volume element/measure $d\mu$ and a real-valued function $\rho$ on $\mathcal{M}$ such that—regarded as a linear operator on the complex Hilbert space $L^2(\mathcal{M}, d\mu)$—the differential operator $iX + \rho$ is symmetric. However, it can be shown that $iX + \rho$ is a genuine *self-adjoint* operator if and only if, as a vector field, $X$ is complete (Abraham & Marsden 1980). This is highly relevant in discussions of the quantisation of a classical mechanical system whose configuration space is a non-trivial differential manifold, as—for example—in a system with constraints, or in a gauge theory. □

**Examples**

1. The vector field $d/dx$ is complete on $\mathbb{R}$ but not on the open submanifold $\mathbb{R}_+$ of positive real numbers. In wave mechanics, this corresponds to the well-known difficulty in defining a self-adjoint momentum operator $\widehat{p}$ in the situation where there is an infinite potential barrier at $x = 0$ that confines the particle to the positive real axis. This is closely related to the fact that for any tolerably well-behaved pair of self-adjoint operators $\widehat{x}, \widehat{p}$ that satisfy the canonical commutation relation

$$[\widehat{x}, \widehat{p}] = i \tag{3.2.9}$$

it can be shown that the eigenvalue spectra of both operators must be the whole real line.

2. On the other hand, the vector field $xd/dx$ *is* complete on $\mathbb{R}_+$. In the problem of quantising a system whose classical configuration space is $\mathbb{R}_+$, this means that it is better to use the observables $x$ and $xp$ rather than the usual pair $x$, $p$ that are appropriate for quantum

mechanics on $\mathbb{R}$. If $\pi := xp$ then the analogue of the CCR Eq. (3.2.9) is the so-called 'affine commutation relation'

$$[\hat{x}, \hat{\pi}] = i\hat{x} \tag{3.2.10}$$

for which, unlike Eq. (3.2.9), genuine self-adjoint operators *can* be found with the spectrum of $\hat{x}$ being just $\mathbb{R}_+$ as required.

This particular example gives just the vaguest hint of how the usual quantum mechanical rules might need to be amended when dealing with a situation in which the classical configuration space is nontrivial. In the general case there is no natural analogue at all of the conventional CCR of Eq. (3.2.9), which is closely locked to systems whose configuration and phase spaces are simple vector spaces. However, whenever there *is* some generalization of Eq. (3.2.9) it is almost invariably a more complicated version of Eq. (3.2.10) in the sense that a canonical operator (*i.e.*, not the unit operator) appears on the right hand side.[2] □

### 3.2.2 One-parameter groups of diffeomorphisms

We turn now to consider the sense in which a vector field, complete or otherwise, can be regarded as the generator of 'infinitesimal' transformations on the manifold. The appropriate starting point is the following definition:

**Definition 3.5**

A *local, one-parameter group of local diffeomorphisms* at a point $p$ in $\mathcal{M}$ consists of (see Figure 3.4):

(i) an open neighbourhood $U$ of $p$;

(ii) a real constant $\epsilon > 0$;

(iii) a family $\{\phi_t \mid |t| < \epsilon\}$ of diffeomorphisms of $U$ onto the open set $\phi_t(U) \subset \mathcal{M}$;

with the following properties:

[2] For an extensive discussion see Isham (1984).

(i) The map

$$\begin{aligned} (-\epsilon, \epsilon) \times U &\to \mathcal{M} \\ (t, q) &\mapsto \phi_t(q) \end{aligned} \tag{3.2.11}$$

is a smooth function of both $t$ and $q$.

(ii) For all $t, s \in \mathbb{R}$ such that $|t|$, $|s|$ and $|t+s|$ are bounded above by $\epsilon$, and for all points $q \in U$ such that $\phi_t(q)$, $\phi_s(q)$ and $\phi_{t+s}(q)$ also belong to $U$, we have

$$\phi_s(\phi_t(q)) = \phi_{s+t}(q). \tag{3.2.12}$$

(iii) For all $q \in U$,

$$\phi_0(q) = q. \tag{3.2.13}$$

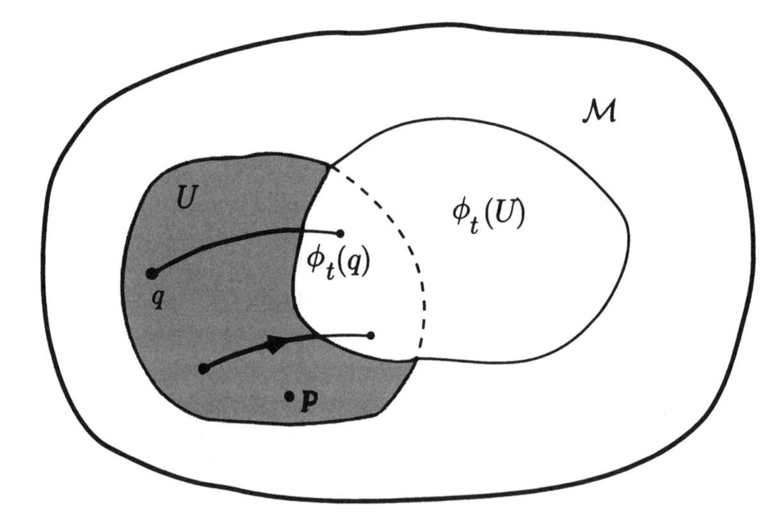


Figure 3.4: A local, one-parameter group of local diffeomorphisms.

## Comments

1. The first occurrence of the word 'local' in the definition refers to the fact that the diffeomorphisms $\phi_t$ are only defined for $t$ lying in some small open interval $(-\epsilon, \epsilon)$ of the real line. The second occurrence of the word refers to the fact that the diffeomorphisms are

defined only locally on $\mathcal{M}$, *i.e.*, they are defined only on the (perhaps very 'small') open subset $U \subset \mathcal{M}$.

For each open set $U$, we denote the set of all such local diffeomorphisms by $\mathrm{Diff}_U(\mathcal{M})$, and note that a pair of such diffeomorphisms $\phi_2$, $\phi_1$ can be combined in the form $\phi_2 \circ \phi_1$ only if $\phi_1(U) \subset U$, in which case $\phi_2 \circ \phi_1$ also belongs to $\mathrm{Diff}_U(\mathcal{M})$. Thus $\mathrm{Diff}_U(\mathcal{M})$ is only what is known as a *partial* group.

2. The term 'one-parameter group' in the definition refers to Eq. (3.2.12) which shows that the map $t \mapsto \phi_t$ can be thought of as a representation of part of the additive group of the real line by local diffeomorphisms of $\mathcal{M}$.

3. One may feel instinctively that one should require the map $(-\epsilon, \epsilon) \to \mathrm{Diff}_U(\mathcal{M})$, $t \mapsto \phi_t$, to be 'smooth' in some sense, but to give a proper meaning to this requires making $\mathrm{Diff}_U(\mathcal{M})$ into a differentiable manifold in its own right. However, this would involve subtle questions of infinite-dimensional differential geometry and, to avoid this, a standard—but highly useful—trick has been employed. Namely, in Eq. (3.2.11), the parameterised family of maps $(-\epsilon, \epsilon) \to \mathrm{Diff}_U(\mathcal{M})$, $t \mapsto \phi_t$ has been replaced by the, equivalent, *single* map $\Phi : (-\epsilon, \epsilon) \times U \to \mathcal{M}$, whose domain and target spaces are both *finite*-dimensional manifolds. There is, therefore, no problem in requiring $\Phi$ to be differentiable, in lieu of the more problematic requirement on $(-\epsilon, \epsilon) \to \mathrm{Diff}_U(\mathcal{M})$.

By this means, we avoid the need to consider issues in infinite-dimensional differential geometry. But, of course, this begs the question of whether requiring the single map $\Phi : (-\epsilon, \epsilon) \times U \to \mathcal{M}$ to be differentiable would be equivalent to requiring the differentiability of $(-\epsilon, \epsilon) \to \mathrm{Diff}_U(\mathcal{M})$ if the latter *could* be made meaningful. It turns out that in analogous problems that *are* well-defined, the answer is "yes", and hence we can have some confidence in the procedure adopted here.

4. Through each point $q \in U$ there passes the local curve $t \mapsto \phi_t(q)$. Since there is such a curve passing through any point in $U$ we can obtain a vector field on $U$ by taking the tangents to this family of curves at each point (see Figure 3.5). The resulting vector field $X^\phi$

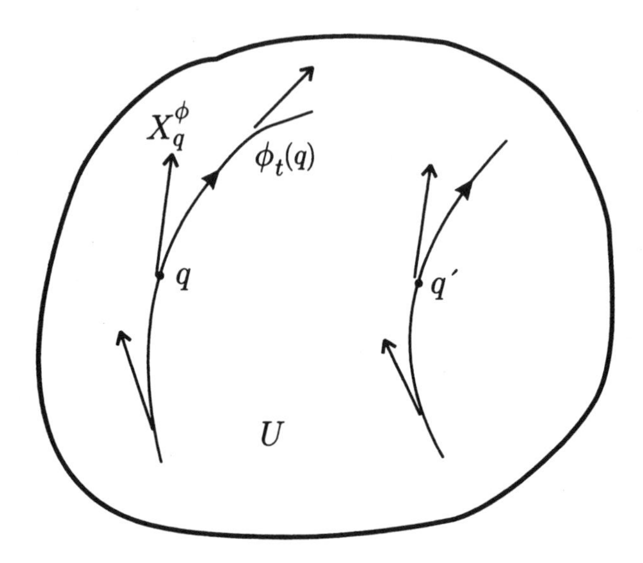


Figure 3.5: The vector field induced by a local one-parameter group.

is said to be *induced* by the family of local diffeomorphisms and is defined as

$$X_q^\phi(f) := \frac{d}{dt} f(\phi_t(q))\bigg|_{t=0} \tag{3.2.14}$$

for all points $q$ in $U \subset \mathcal{M}$. □

The following proposition seems intuitively obvious but is nevertheless worth proving properly:

**Theorem 3.1** *For all points $q \in U$ the curve $t \mapsto \phi_t(q)$ is an integral curve of $X^\phi$ for $|t| < \epsilon$.*

**Proof**

Define the curve $\phi_q : (-\epsilon, \epsilon) \to \mathcal{M}$ by $\phi_q(t) := \phi_t(q)$. According to the definition of an integral curve in Eq. (3.2.3) we want to show that

$$\phi_{q*}\left(\frac{d}{dt}\right)_s = X^\phi_{\phi_s(q)} \text{ for } |s| < \epsilon. \tag{3.2.15}$$

But

$$\phi_{q*}\left(\frac{d}{dt}\right)_s f = \frac{d}{dt}f(\phi_q(t))\bigg|_{t=s} = \frac{d}{dt}f(\phi_t(q))\bigg|_{t=s}. \tag{3.2.16}$$

Let $t = s + u$. Then the right hand side of Eq. (3.2.16) is

$$\begin{aligned}\frac{d}{du}f(\phi_{s+u}(q))\bigg|_{u=0} &= \frac{d}{du}f(\phi_u(\phi_s(q)))\bigg|_{u=0} \\ &= X^{\phi}_{\phi_s(q)}f. \end{aligned} \tag{3.2.17}$$

### 3.2.3 Local flows

We have seen how to go from a local group of local diffeomorphisms to a vector field. But the problem of interpreting an equation like Eq. (3.2.1) involves the converse procedure, *i.e.*, going from a vector field to a group of diffeomorphisms. The central concept here is of a 'local flow':

**Definition 3.6**

> Let $X$ be a vector field defined on an open subset $U$ of a manifold $\mathcal{M}$ and let $p$ be a point in $U \subset \mathcal{M}$. Then a *local flow* of $X$ at $p$ is a local one-parameter group of local diffeomorphisms defined on some open subset $V$ of $U$ such that $p \in V \subset U$ and such that the vector field induced by this family is equal to the given field $X$.

**Comments**

1. Local flows always exist, and are unique, by virtue of the usual existence and uniqueness theorems of ordinary differential equation theory.[3]

2. If $X$ is a complete vector field then $V$ can always be chosen to be the entire manifold $\mathcal{M}$, and the interval $(-\epsilon, \epsilon)$ can be extended

[3]For a proof of this and related analytical theorems on vector fields see Abraham & Marsden (1980).

to the entire real line. In this case, if $\phi_t^X$ denotes the family of diffeomorphisms that is the flow for $X$, then the map $t \mapsto \phi_t^X$ is an isomorphism from the entire additive group of the real line into the group Diff($\mathcal{M}$) of diffeomorphisms of $\mathcal{M}$. We can say that this defines a one-parameter subgroup of Diff($\mathcal{M}$) and hence an element of its Lie algebra (see later for a full discussion of the relation between Lie groups and Lie algebras). This is the sense in which a vector field is said to belong to the Lie algebra of Diff($\mathcal{M}$).

3. In a local coordinate system, the local flow $\phi_t^X$ satisfies the differential equation (*cf.* Eq. (3.2.6))

$$X^\mu(\phi_t^X(q)) = \frac{d}{dt}x^\mu(\phi_t^X(q)) \tag{3.2.18}$$

around any point $q$ in the open set $U \subset \mathcal{M}$. A Taylor expansion of the coordinates $x^\mu(\phi_t^X(q))$ of the transformed point $\phi_t^X(q)$ around the value $t = 0$ gives

$$x^\mu(\phi_t^X(q)) = x^\mu(q) + tX^\mu(q) + O(t^2) \tag{3.2.19}$$

which can be regarded as a reasonably rigorous version of the heuristic Eq. (3.2.1) and the claim that a vector field generates 'infinitesimal transformations'.

4. If $X$ and $Y$ are two vector fields on $\mathcal{M}$ it can be shown that [Exercise!]

$$\left.\frac{d}{dt}\phi^X_{-t*}(Y)\right|_{t=0} = [X, Y] \tag{3.2.20}$$

where

$$\left.\frac{d}{dt}\phi^X_{-t*}(Y)\right|_{t=0}(p) := \lim_{\epsilon\to 0}\frac{\phi^X_{-\epsilon*}Y_{\phi^X_\epsilon(p)} - Y_p}{\epsilon}. \tag{3.2.21}$$

The left hand side of Eq. (3.2.20) is often called the *Lie derivative* of $Y$ with respect to $X$ and is denoted $L_X Y$. It is a measure of the extent to which the vector field $Y$ changes as it is 'dragged' along the integral curves of $X$.

### 3.2.4 Some concrete examples of integral curves and flows

The ideas of integral curves and flows can be usefully illustrated with the aid of some examples of a vector field on a simple vector space manifold.

**Problem 1.**

(i) Find the integral curves of the vector fields $ay\frac{\partial}{\partial x} + bx\frac{\partial}{\partial y}$ on the manifold $\mathbb{R}^2$, where $a$ and $b$ are an arbitrary pair of real numbers. Are these fields complete?

(ii) Using these results, compute the flow $\phi_t^X$ of the vector field $X = y\frac{\partial}{\partial x}$, and show by explicit calculation that, if $Y := x\frac{\partial}{\partial y}$, the commutator $[Y, X]$ satisfies the relation $[Y, X] = \lim_{t\to 0} \frac{(\phi_{t*}^X Y - Y)}{t}$.

**Answer**

(i) In the usual global Cartesian coordinates $(x, y)$ on $\mathbb{R}^2$, the integral curves $t \mapsto (\sigma^x(t), \sigma^y(t))$ of the vector field $X := ay\frac{\partial}{\partial x} + bx\frac{\partial}{\partial y}$ satisfy the differential equations

$$\frac{d\sigma^x}{dt}(t) = X^x(\sigma^x(t), \sigma^y(t)) = a\sigma^y(t) \tag{3.2.22}$$

$$\frac{d\sigma^y}{dt}(t) = X^y(\sigma^x(t), \sigma^y(t)) = b\sigma^x(t), \tag{3.2.23}$$

where $X^x$ and $X^y$ are the components of $X$ in this coordinate system, and where $\sigma^\mu(t) := x^\mu \circ \sigma(t)$. Differentiating Eq. (3.2.22) with respect to $t$ and substituting Eq. (3.2.23) in the result gives

$$\frac{d^2\sigma^x}{dt^2}(t) = a\frac{d\sigma^y}{dt}(t) = ab\,\sigma^x(t), \tag{3.2.24}$$

and hence there are three types of integral curve, depending on the values of the real numbers $a$ and $b$:

**(1)** $b = 0$: The integral curve passing through the point $(x_0, y_0)$ (*i.e.*, $x_0 = \sigma^x(0)$ and $y_0 = \sigma^y(0)$) is $\sigma^y(t) = y_0$, $\sigma^x(t) = x_0 + ay_0t$.

Similarly, if $a$=0, the integral curves are $\sigma^x(t) = x_0$, $\sigma^y(t) = y_0 + bx_0t$.

**(2)** $ab > 0$: The solution to the differential equations Eqs. (3.2.22–3.2.23) is now

$$\sigma^x(t) = A\cosh\omega t + B\sinh\omega t \tag{3.2.25}$$
$$\sigma^y(t) = b(A\sinh\omega t - B\cosh\omega t) \tag{3.2.26}$$

where $\omega := \sqrt{ab}$, and $A$ and $B$ are integration constants. Clearly, $A = \sigma^x(0)$ and $B = -\sigma^y(0)/b$.

**(3)** $ab < 0$: The solution to Eqs. (3.2.22–3.2.23) is now

$$\sigma^x(t) = A\cos\omega t + B\sin\omega t \tag{3.2.27}$$
$$\sigma^y(t) = b(A\sin\omega t - B\cos\omega t) \tag{3.2.28}$$

where $\omega := \sqrt{-ab}$. Once again, $A = \sigma^x(0)$ and $B = -\sigma^y(0)/b$.

We note that the vector fields are complete in all three cases since the solutions can all be defined for an arbitrarily large value of the parameter $t$.

(ii) The integral curves for the vector field $X = y\frac{\partial}{\partial x}$ are $\sigma^x(t) = x_0 + y_0t$, $\sigma^y(t) = y_0$. Thus the flow $\phi_t^X$ of $X$ is the one-parameter family of diffeomorphisms of $\mathbb{R}^2$ given by

$$\phi_t^X(c, d) = (c + dt, d) \tag{3.2.29}$$

for all points $(c, d) \in \mathbb{R}^2$.

To compute the desired limit, it is necessary to work out the components of the vector fields $\phi_{t\,*}^X Y$ and $Y$, and subtract their values at a general point $(c, d) \in \mathbb{R}^2$. Thus we must subtract $Y_{(c,d)}$ from $\phi_{t\,*}^X\left(Y_{\phi^X{}_t^{-1}(c,d)}\right)$. As usual, the components $Z^\mu$ of a vector field $Z$ are equal to $Z(x^\mu)$, and hence, using the fact that $\phi^{X}{}_t^{-1}(c,d) = (c-dt, d)$, we get

$$\left(\phi^X_t{}_* Y_{\phi^X{}_t^{-1}(c,d)}\right)^{\mu=x} = (c-dt)\left(\frac{\partial}{\partial y}\right)_{(c-dt,d)}(x+yt) \quad (3.2.30)$$

$$= (c-dt)t \quad (3.2.31)$$

$$\left(\phi^X_t{}_* Y_{\phi^X{}_t^{-1}(c,d)}\right)^{\mu=y} = (c-dt)\left(\frac{\partial}{\partial y}\right)_{(c-dt,d)} y \quad (3.2.32)$$

$$= (c-dt) \quad (3.2.33)$$

where we have also used the result $(\phi_* Y_p)(f) = Y_p(f\circ\phi)$. Hence

$$(\phi^X_t{}_* Y)_{(c,d)} = \phi^X_t{}_*\left(Y_{\phi^X{}_t^{-1}(c,d)}\right) = (c-dt)\left(t\frac{\partial}{\partial x}+\frac{\partial}{\partial y}\right)_{(c,d)} \quad (3.2.34)$$

and thus, for an arbitrary point $(c,d)\in\mathbb{R}^2$,

$$\frac{(\phi^X_t{}_* Y - Y)_{(c,d)}}{t} = \left(c\frac{\partial}{\partial x} - d\frac{\partial}{\partial y}\right)_{(c,d)} - dt\left(\frac{\partial}{\partial x}\right)_{(c,d)}. \quad (3.2.35)$$

Hence, for each $t\in\mathbb{R}$, the vector fields $\phi^X_t{}_* Y$ and $Y$ satisfy the equation

$$\frac{(\phi^X_t{}_* Y - Y)}{t} = \left(x\frac{\partial}{\partial x} - y\frac{\partial}{\partial y}\right) - ty\frac{\partial}{\partial x}, \quad (3.2.36)$$

and therefore

$$\lim_{t\to 0}\frac{(\phi^X_t{}_* Y - Y)}{t} = x\frac{\partial}{\partial x} - y\frac{\partial}{\partial y}. \quad (3.2.37)$$

However

$$[Y,X] = x\frac{\partial}{\partial y}\left(y\frac{\partial}{\partial x}\right) - y\frac{\partial}{\partial x}\left(x\frac{\partial}{\partial y}\right) \quad (3.2.38)$$

$$= x\frac{\partial}{\partial x} - y\frac{\partial}{\partial y}, \quad (3.2.39)$$

which proves the result. **QED**

**Problem 2.**

Let $X := y^2\frac{\partial}{\partial x}$ and $Y := x^2\frac{\partial}{\partial y}$ be a pair of vector fields on $\mathbb{R}^2$. Show that they are complete but that $X + Y$ is incomplete.

**Answer**

(i) In the usual global Cartesian coordinates $(x, y)$ on $\mathbb{R}^2$, the integral curves $t \mapsto (\sigma^x(t), \sigma^y(t))$ of the vector field $X := y^2\frac{\partial}{\partial x}$ satisfy the differential equations

$$\frac{d\sigma^x}{dt}(t) = X^x(\sigma^x(t), \sigma^y(t)) = (\sigma^y(t))^2 \tag{3.2.40}$$

$$\frac{d\sigma^y}{dt}(t) = X^y(\sigma^x(t), \sigma^y(t)) = 0 \tag{3.2.41}$$

from which it follows at once that the integral curve of $X$ that passes through the point $(x_0, y_0) \in \mathbb{R}^2$ is $\sigma^y(t) = y_0$, $\sigma^x(t) = x_0 + ty_0^2$. Clearly this curve is defined for any value of $t \in \mathbb{R}$, and hence the vector field $X$ is complete.

(ii) The same analysis, but with $x$ and $y$ interchanged, shows that the integral curve of $Y := x^2\frac{\partial}{\partial y}$ is $\sigma^x(t) = x_0$, $\sigma^y(t) = y_0 + tx_0^2$. Thus $Y$ too is complete.

(iii) The integral curves of $X + Y = y^2\frac{\partial}{\partial x} + x^2\frac{\partial}{\partial y}$ satisfy the differential equations

$$\frac{d\sigma^x}{dt}(t) = (X^x + Y^x)(\sigma^x(t), \sigma^y(t)) = (\sigma^y(t))^2 \tag{3.2.42}$$

$$\frac{d\sigma^y}{dt}(t) = (X^y + Y^y)(\sigma^x(t), \sigma^y(t)) = (\sigma^x(t))^2 \tag{3.2.43}$$

which are not easy to solve explicitly. However, by inspection, it is clear that the equations are consistent with setting $\sigma^x(t) = \sigma^y(t)$, and then the solutions of the single ensuing differential equation $d\sigma^x(t)/dt = (\sigma^x(t))^2$ are

$$\sigma^x(t) = \sigma^y(t) = \frac{1}{\tau - t} \tag{3.2.44}$$

for any $\tau \in \mathbb{R}$. Thus Eq. (3.2.44) is the integral curve of $X + Y$ that passes through the point $(1/\tau, 1/\tau)$. However, unlike $X$ and $Y$ individually, the sum $X + Y$ is *not* complete since the integral curve Eq. (3.2.44) has a singularity at $t = \tau$, and hence cannot be defined for all $t \in \mathbb{R}$.

# 3.3 Cotangent Vectors

## 3.3.1 The algebraic dual of a vector space

For any real vector space $V$ there is an associated 'dual' space of all real-valued linear maps on $V$. When applied to the vector spaces $T_p\mathcal{M}$, $p \in \mathcal{M}$, this well-known construction turns out to be of the greatest value. First we briefly recall some of the principle ideas in the theory of dual spaces.

**Definition 3.7**

1. If $V$ is a real vector space, the (algebraic) *dual* of $V$ is defined to be the set $V^*$ of all linear maps $L : V \to \mathbb{R}$.

   The value of $L \in V^*$ on $v$ is conventionally written as $\langle L, v \rangle$, or as $\langle L, v \rangle_V$ if it is necessary to emphasise the vector space involved.

2. The set $V^*$ can be given the structure of a real vector space by defining
$$\langle L_1 + L_2, v \rangle := \langle L_1, v \rangle + \langle L_2, v \rangle \tag{3.3.1}$$
   for all $L_1, L_2 \in V^*$, $v \in V$; and
$$\langle rL, v \rangle := r\langle L, v \rangle \tag{3.3.2}$$
   for all $r \in \mathbb{R}$, $L \in V^*$ and $v \in V$.

3. If $V$ has a finite dimension $n$, and if $\{e_1, e_2, \ldots e_n\}$ is a basis for $V$, the *dual basis* for the vector space $V^*$ is the set of vectors $\{f^1, f^2, \ldots, f^n\}$ in $V^*$ that are determined uniquely by the requirement
$$\langle f^i, e_j \rangle = \delta^i_j \tag{3.3.3}$$
   for all $i, j = 1 \ldots n$.

4. If $L : V \to W$ is a linear map from a vector space $V$ to a vector space $W$, the *dual* $L^* : W^* \to V^*$ of $L$ is defined on $k \in W^*$ by
$$\langle L^*k, v \rangle_V := \langle k, Lv \rangle_W, \tag{3.3.4}$$

for all $v \in V$. We note that the map $L^* : W^* \to V^*$ is linear with respect to the vector space structures on the dual spaces $W^*$ and $V^*$.

### Comments

1. The existence of the dual basis defined by Eq. (3.3.3) means that $\dim V = \dim V^*$, and hence $V$ and $V^*$ are isomorphic as vector spaces.

This isomorphism is non-canonical (in the sense that there is no basis-independent isomorphism) but, nonetheless, one may wonder what the point is of constructing $V^*$ if it is isomorphic to the starting space $V$. As we shall see shortly, in the context of differential geometry with $V = T_p\mathcal{M}$ the answer lies in the existence of the pull-back map defined in Eq. (3.3.4).

2. For any vector space $V$ there is a canonical embedding $\chi : V \to (V^*)^*$ defined by

$$\langle \chi(v), \ell \rangle_{V^*} := \langle \ell, v \rangle_V \tag{3.3.5}$$

for all $v \in V$ and $\ell \in V^*$. If $V$ has a finite dimension, then the map $\chi$ in Eq. (3.3.5) is an isomorphism.

3. The question of the dual space is quite different if $V$ has an infinite dimension. In this case, it is important to take explicit note of the topology on $V$, which is invariably chosen so that $V$ is a topological vector space, *i.e.*, the addition of two vectors and the multiplication of a vector by a scalar are required to be continuous operations. This requirement is relatively unimportant if $V$ has a finite dimension since, as mentioned earlier, there is a unique Hausdorff topology on $V$ for which it becomes a topological vector space. Furthermore, any linear map from $V$ to $\mathbb{R}$ is necessarily continuous.

However, in the infinite-dimensional case there are typically many linear maps from $V$ to $\mathbb{R}$ that are *not* continuous, and hence it is necessary to distinguish between the *algebraic dual* $V^*$—defined to be the set of all linear maps from $V$ to $\mathbb{R}$—and the *topological dual* $V'$—defined to be the set of all *continuous* linear maps from $V$ to $\mathbb{R}$. The existence of this important distinction adds much complexity to the discussion, and is one of the reasons why the development of infinite-dimensional differential geometry requires the use of considerably more sophisticated tools drawn from functional analysis than does the finite-dimensional case. □

### 3.3.2 The main definitions

We come now to the topic of cotangent vectors and differential forms—ideas that are of the greatest importance in the mathematical development of differential geometry and in its applications to theoretical physics.

**Definition 3.8**

1. A *cotangent vector* at a point $p$ in a manifold $\mathcal{M}$ is a real linear map $k$ from $T_p\mathcal{M}$ into $\mathbb{R}$. The value of $k$ on the tangent vector $v \in T_p\mathcal{M}$ will be written as $\langle k, v\rangle$, or $\langle k, v\rangle_p$ if it is necessary to emphasise the point in the manifold with which we are concerned.

2. The *cotangent space* at $p \in \mathcal{M}$ is the set $T_p^*\mathcal{M}$ of all such linear maps, *i.e.*, it is the dual of the vector space $T_p\mathcal{M}$. It follows from the discussion above that (i) this dual space is itself a vector space; and (ii) $\dim T_p^*\mathcal{M} = \dim T_p\mathcal{M}$ $[= \dim \mathcal{M}]$.

3. The *cotangent bundle* $T^*\mathcal{M}$ is the set of all cotangent vectors at all points in $\mathcal{M}$:

$$T^*\mathcal{M} := \bigcup_{p\in\mathcal{M}} T_p^*\mathcal{M}. \tag{3.3.6}$$

   It is a vector bundle and, as in the case of the tangent bundle $T\mathcal{M}$, it can be equipped with a natural structure of a $2m$-dimensional differentiable manifold; see below (*cf.* Section 2.3.5).

4. A *one-form* $\omega$ on $\mathcal{M}$ is a smooth assignment of a cotangent vector $\omega_p$ to each point $p \in \mathcal{M}$. In this context, 'smooth' means

that for any vector field $X$ on $\mathcal{M}$ the real-valued function on $\mathcal{M}$

$$\langle\omega, X\rangle(p) := \langle\omega_p, X_p\rangle_p \tag{3.3.7}$$

is smooth.

Equivalently, a one-form can be defined as a smooth cross-section of the cotangent bundle (see below).

## Comments

1. A very important application of these ideas is to classical Hamiltonian dynamics. If such a classical mechanical system has a configuration space $Q$ then the space of classical states is defined to be the cotangent bundle $T^*Q$.

2. Care needs to be taken when interpreting Eq. (3.3.7). It might perhaps be construed as asserting that a one-form is a dual of the space of vector fields in the same sense that a cotangent vector is an element of the dual space of a space $T_p\mathcal{M}$ of tangent vectors. However, this is certainly *not* the case if the set of vector fields is regarded as a vector space over $\mathbb{R}$, since $\langle\omega, X\rangle$ as defined in Eq. (3.3.7) is a function on $\mathcal{M}$, not a real number.

The analogy becomes closer if the vector fields are regarded instead as a module over the ring $C^\infty(\mathcal{M})$, but even then care is needed as these are infinite-dimensional spaces and topological subtleties arise in discussions of the precise structure of the dual of such a space. For our purposes, it is safest just to accept (3.3.7) as it stands without trying to reinterpret it in the full language of duals of topological vector spaces.

3. As was shown in Section 2.3.5, any local coordinate system around a point $p \in \mathcal{M}$ generates a basis set

$$\left(\frac{\partial}{\partial x^1}\right)_p, \ldots, \left(\frac{\partial}{\partial x^m}\right)_p \tag{3.3.8}$$

of derivations for the tangent space $T_p\mathcal{M} \cong D_p\mathcal{M}$. As usual in linear algebra, and as discussed above, there is an associated basis for the dual space $T_p^*\mathcal{M}$. This is denoted

$$(dx^1)_p, (dx^2)_p, \ldots, (dx^m)_p \tag{3.3.9}$$

and is defined by (*cf.* Eq. (3.3.3))

$$\langle (dx^\mu)_p, \left(\frac{\partial}{\partial x^\nu}\right)_p \rangle_p := \delta^\mu_\nu \tag{3.3.10}$$

for all $\mu, \nu = 1, 2, \ldots, m$.

Thus any $k \in T_p^*\mathcal{M}$ can be expanded as

$$k = \sum_{\mu=1}^{m} k_\mu (dx^\mu)_p \tag{3.3.11}$$

where the *components* of $k$ with respect to the given coordinate system are the set of real numbers $k_\mu$ satisfying

$$k_\mu = \langle k, \left(\frac{\partial}{\partial x^\mu}\right)_p \rangle_p \tag{3.3.12}$$

where $\mu = 1, 2, \ldots, m$. It also follows that the pairing $\langle k, v\rangle$ can be written in terms of these local coordinate components as

$$\langle k, v\rangle = \sum_{\mu=1}^{m} k_\mu v^\mu \tag{3.3.13}$$

where (see Eq. (2.3.30)) $v^\mu = v(x^\mu)$.

4. These remarks extend to a local coordinate representation for one-forms, and we write

$$\omega = \sum_{\mu=1}^{m} \omega_\mu \, dx^\mu \tag{3.3.14}$$

which is short hand for

$$\omega_p = \sum_{\mu=1}^{m} \omega_\mu(p)(dx^\mu)_p \tag{3.3.15}$$

for all $p$ in the domain of the coordinate chart. Here the components $\omega_\mu$ of the one-form $\omega$ are the functions defined on the domain $U$ of the coordinate chart by

$$\omega_\mu(q) := \langle \omega, \left(\frac{\partial}{\partial x^\mu}\right)_q \rangle_q \tag{3.3.16}$$

for all $q \in U \subset \mathcal{M}$.

5. The cotangent bundle $T^*\mathcal{M}$ can be given the structure of a $2m$-dimensional differentiable manifold in a way that closely parallels the manner in which this was done for the tangent bundle $T\mathcal{M}$. More precisely, if $(U, \phi)$ is a local coordinate chart on $\mathcal{M}$ then the associated $2m$ coordinates of a cotangent vector $k$ at any point $p$ in $U$ are defined to be (i) the $m$ real numbers $\{x^1(p), x^2(p), \ldots, x^m(p)\}$; and (ii) the $m$ real numbers $\{k_1, k_2, \ldots, k_m\}$ where $k_\mu := \langle k, \left(\frac{\partial}{\partial x^\mu}\right)_p \rangle_p$.

This construction provides an alternative way of defining a one-form: namely, as a smooth cross-section of the cotangent bundle $T^*\mathcal{M}$. Clearly this is analogous to the way in which a vector field can be defined as a smooth cross-section of the tangent bundle $T\mathcal{M}$.

6. There are some situations in which it is useful to generalise the notion of a cotangent vector to allow for linear maps from $T_p\mathcal{M}$ to a general real vector space $V$, not just the real numbers. A one-form of this type is called a *$V$-valued one-form*. We shall see an important example of this idea in Section 4.3.2 in the context of Lie group theory. □

### 3.3.3 The pull-back of a one-form

Now we come to a most important property of one-forms. Namely, even though it is not true in general that a map $h : \mathcal{M} \to \mathcal{N}$ between two manifolds $\mathcal{M}$ and $\mathcal{N}$ can be used to 'push forward' a vector field on $\mathcal{M}$, it *can* be used to 'pull back' a one-form on $\mathcal{N}$. This feature of forms is of the utmost significance; it is deeply connected with the way in which global topological structure of a manifold are reflected in the DeRham cohomology groups of $\mathcal{M}$ defined using differential forms. It is also closely linked with the fact that fibre bundles also 'pull back', not 'push forward' (see later).

The first step is to define the 'pull-back' map $h^* : T^*_{h(p)}\mathcal{N} \to T^*_p\mathcal{M}$ on the individual cotangent spaces using the construction defined in Eq. (3.3.4).

**Definition 3.9**

1. If $h : \mathcal{M} \to \mathcal{N}$ we have the linear map $h_* : T_p\mathcal{M} \to T_{h(p)}\mathcal{N}$, and then the map $h^* : T^*_{h(p)}\mathcal{N} \to T^*_p\mathcal{M}$ is defined as the dual of $h_*$. Thus, for all $k \in T^*_{h(p)}\mathcal{N}$ and $v \in T_p\mathcal{M}$, (*cf.* Eq. (3.3.4)),

$$\langle h^*k, v\rangle_p := \langle k, h_*v\rangle_{h(p)} \tag{3.3.17}$$

as illustrated in Figure 3.6.

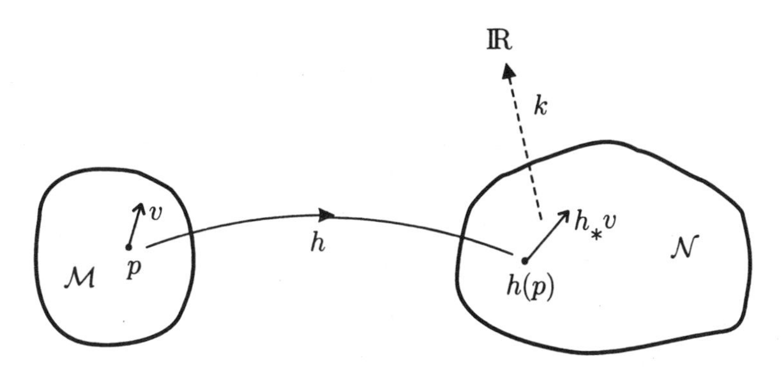


Figure 3.6: The definition of the pull-back $h^*k$.

2. It is matter of the greatest importance that this definition can be extended to one-forms. Specifically, if $\omega$ is a one-form on $\mathcal{N}$ then the *pull-back* of $\omega$ is the one-form $h^*\omega$ on $\mathcal{M}$ defined by

$$\langle h^*\omega, v\rangle_p := \langle \omega, h_*v\rangle_{h(p)} \tag{3.3.18}$$

for all points $p \in \mathcal{M}$ and all tangent vectors $v \in T_p\mathcal{M}$.

## Comments

1. If $\mathcal{M}$, $\mathcal{N}$ and $\mathcal{P}$ are three differentiable manifolds with maps $h : \mathcal{M} \to \mathcal{N}$ and $k : \mathcal{N} \to \mathcal{P}$, then (*cf.* Eq. (2.3.11))

$$(k \circ h)^* = h^* k^*. \tag{3.3.19}$$

2. The pull-back map $h^*$ is only defined on those cotangent spaces $T^*_q\mathcal{N}$ for which the point $q \in \mathcal{N}$ is such that there exists $p \in \mathcal{M}$ with $q = h(p)$. It is therefore not necessarily correct to think of $h^*$ as inducing a map from the complete cotangent bundle $T^*\mathcal{N}$ into $T^*\mathcal{M}$: this will only be the case if $h$ is a *surjective* map from $\mathcal{M}$ to $\mathcal{N}$.

3. There is an analogue for the pull-back of a one-form of the expression Eq. (2.3.46) of the push-forward of a tangent vector expressed in local coordinates on the domain and target manifolds. Specifically, if $h : \mathcal{M} \to \mathcal{N}$, let $\{x^1, x^2, \ldots, x^n\}$ and $\{y^1, y^2, \ldots, y^m\}$ be local coordinates on $\mathcal{M}$ and $\mathcal{N}$ respectively. Then if $\omega$ is a one-form on $\mathcal{N}$, we have the local coordinate representation

$$\omega_{h(p)} = \sum_{\nu=1}^{n} \omega_\nu(h(p))\,(dy^\nu)_{h(p)} \tag{3.3.20}$$

for all points $p \in \mathcal{M}$ such that $h(p)$ lies in the domain of the $y$-coordinate system on $\mathcal{N}$.

If, in addition, $p$ lies in the domain of the $x$-coordinate system on $\mathcal{M}$, then the components $(h^*\omega)_\mu$ with respect to this coordinate system of the pull-back $h^*\omega$ are

$$(h^*\omega)_\mu(p) = \langle h^*\omega, (\partial/\partial x^\mu)\rangle_p := \langle \omega, h_*(\partial/\partial x^\mu)_p\rangle_{h(p)} \tag{3.3.21}$$

$$= \sum_{\nu=1}^{n} \omega_\nu(h(p))\,(h_*(\partial/\partial x^\mu)_p)^\nu \tag{3.3.22}$$

where we have used the definition Eq. (3.3.18) of the pull-back operation and Eq. (3.3.13). However, we saw earlier in Eq. (2.3.44) that the components $(h_*(\partial/\partial x^\mu)_p)^\nu$ of the push-forward of the derivation $(\partial/\partial x^\mu)_p$ at the point $p$, can be expressed in terms of the Jacobian matrix of the map $h$. Inserting this into Eq. (3.3.22), gives the result

$$(h^*\omega)_p = \sum_{\nu=1}^{n}\sum_{\mu=1}^{m} \omega_\nu(h(p))\,\frac{\partial h^\nu}{\partial x^\mu}(p)\,(dx^\mu)_p, \tag{3.3.23}$$

which is the analogue for pull-backs of the local representation of a push-forward in Eq. (2.3.46). □

### 3.3.4 A simple example of the pull-back operation

Let us illustrate the pull-back operation with the aid of the same map $h : \mathbb{R}^2 \to \mathbb{R}^2$ that was used in Section 2.3.7 to illustrate the push-forward of a tangent vector.

**Problem**

The map $h : \mathbb{R}^2 \to \mathbb{R}^2$ is defined by

$$h(a,b) := (a^2 - 2b, 4a^3b^2). \tag{3.3.24}$$

Compute the pull-back by $h$ of the one-form $\omega$ defined in global coordinates $\{y^1, y^2\}$ on the target manifold $\mathbb{R}^2$ by

$$\omega_{(c,d)} := cd(dy^1)_{(c,d)} + c^2(dy^2)_{(c,d)} \tag{3.3.25}$$

for all points $(c,d) \in \mathbb{R}^2$.

**Answer**

According to Eq. (3.3.23), the components of the pull-back $h^*\omega$ with respect to the global coordinate system $\{x^1, x^2\}$ on the domain manifold $\mathbb{R}^2$ are

$$(h^*\omega)_1(a,b) = \omega_1(h(a,b))\frac{\partial h^1}{\partial x^1}(a,b) + \omega_2(h(a,b))\frac{\partial h^2}{\partial x^1}(a,b) \tag{3.3.26}$$

$$(h^*\omega)_2(a,b) = \omega_1(h(a,b))\frac{\partial h^1}{\partial x^2}(a,b) + \omega_2(h(a,b))\frac{\partial h^2}{\partial x^2}(a,b) \tag{3.3.27}$$

at any point $(a,b)$ in the domain manifold $\mathbb{R}^2$. Using the explicit form of $h$, and the expressions in Eq. (2.3.60) for the Jacobian matrix of the map $h$, we get

$$\begin{aligned}(h^*\omega)_1(a,b) &= \omega_1(a^2 - 2b, 4a^3b^2)2a + \omega_2(a^2 - 2b, 4a^3b^2)12a^2b^2 \\ &= (a^2 - 2b)4a^3b^2 \times 2a + (a^2 - 2b)^2 12a^2b^2\end{aligned}$$

$$
\begin{aligned}
&= 4a^2b^2(5a^4 - 16a^2b + 12b^2) && (3.3.28)\\
(h^*\omega)_2(a,b) &= \omega_1(a^2-2b, 4a^3b^2)(-2) + \omega_2(a^2-2b, 4a^3b^2)8a^3b \\
&= (a^2-2b)4a^3b^2 \times (-2) + (a^2-2b)^2 8a^3b \\
&= 8a^3b(a^4 - 5a^2b + 6b^2). && (3.3.29)
\end{aligned}
$$

Hence, the final result is

$$
\begin{aligned}
(h^*\omega)_{(a,b)} = {}& 4a^2b^2(5a^4 - 16a^2b + 12b^2)(dx^1)_{(a,b)} \\
&+8a^3b(a^4 - 5a^2b + 6b^2)(dx^2)_{(a,b)} \quad (3.3.30)
\end{aligned}
$$

at any point $(a, b)$ in the domain manifold $\mathbb{R}^2$.

### 3.3.5 The Lie derivative

An important special case of the pull-back operation on a one-form $\omega$ on a manifold $\mathcal{M}$ occurs in the context of a vector field $X$ with an associated local, one-parameter group of local diffeomorphisms $t \to \phi_t^X$. For each $t \in (-\epsilon, \epsilon)$, we have the pull-back one-form $\phi_t^{X*}(\omega)$ (also on $\mathcal{M}$), and this one-parameter family of forms can be used to describe the way in which $\omega$ changes along the flow lines of $X$. More precisely, we have the following definition.

**Definition 3.10**

The *Lie derivative* $L_X\omega$ of a one-form $\omega$ with respect to a vector field $X$ is defined to be the rate of change of $\omega$ along the flow lines of the one-parameter group $\phi_t^X$ of local diffeomorphisms associated with $X$:

$$L_X\omega := \frac{d}{dt}\phi_t^{X*}\omega\Big|_{t=0} \qquad (3.3.31)$$

where the derivative on the right hand side is defined in analogy to Eq. (3.2.21).

**Comments**

1. It can be shown that [Exercise!]:

(i)

$$\left.\frac{d}{dt}\phi_t^{X*}\omega\right|_{t=s} = \phi^{X*}_s(L_X\omega), \tag{3.3.32}$$

(ii)

$$L_X\langle\omega, Y\rangle = \langle L_X\omega, Y\rangle + \langle\omega, L_XY\rangle \tag{3.3.33}$$

for all vector fields $X$ and $Y$. Recall that (i) on a function $f$, $L_Xf := Xf$; and (ii) $L_XY := [X, Y]$ (see Eq. (3.2.20)).

(iii) In a local coordinate system, the components of the one-form $L_X\omega$ are related to those of $\omega$ and $X$ by

$$(L_X\omega)_\mu = \sum_{\nu=1}^{m}(\omega_{\mu,\nu}X^\nu + \omega_\nu X^\nu{}_{,\mu}). \tag{3.3.34}$$

2. The Lie derivative $L_Xf := Xf$ of a function $f$ along $X$ arises also in the definition of the 'exterior derivative' of $f$—a concept that generalises to a general $n$-form (see later; it is convenient for this reason to think of a function as a '0-form') and which is of the greatest significance:

**Definition 3.11**

The *exterior derivative* of a function $f \in C^\infty(\mathcal{M})$ is the one-form $df$ defined by

$$\langle df, X\rangle := Xf = L_Xf \tag{3.3.35}$$

for all vector fields $X$ on $\mathcal{M}$. In local coordinates:

$$(df)_p = \sum_{\mu=1}^{m}\left(\frac{\partial}{\partial x^\mu}\right)_p f\,(dx^\mu)_p. \tag{3.3.36}$$

□

**Exercises**

Show that exterior differentiation commutes with the pull-back operation in the sense that, if $h : \mathcal{M} \to \mathcal{N}$, then

$$h^*(df) = d(f \circ h) \tag{3.3.37}$$

for all $f \in C^\infty(\mathcal{N})$.

## 3.4 General Tensors and $n$-Forms

### 3.4.1 The tensor product operation

In order to proceed from tangent and cotangent vectors to general tensors and $n$-forms it is necessary to introduce the idea of the tensor product $V \otimes W$ of two vector spaces[4] $V$ and $W$.

The most general and powerful approach is via the 'universal factorisation property' in which a tensor product of $V$ and $W$ is defined to be a vector space, written $V \otimes W$, and a bilinear map $\mu : V \times W \to V \otimes W$ with the property that given any other vector space $Z$ and a bilinear map $b : V \times W \to Z$ there exists a unique linear map $\tilde{b} : V \otimes W \to Z$ such that the following diagram commutes

$$\begin{array}{ccc} V \times W & \xrightarrow{b} & Z \\ \downarrow{\scriptstyle \mu} & \nearrow{\scriptstyle \tilde{b}} & \\ V \otimes W & & \end{array} \tag{3.4.1}$$

It is easy to show that the pair $(V \otimes W, \mu)$ is unique up to isomorphisms.

This definition is very general, but it is also rather abstract, and for our purposes it is useful to exploit the fact that, for finite-dimensional vector spaces, there is a natural isomorphism $j : V \otimes W \to B(V^* \times W^*, \mathbb{R})$ where $B(V^* \times W^*, \mathbb{R})$ is the vector space of bilinear maps from the Cartesian product $V^* \times W^*$ of the duals of $V$ and $W$ into the real numbers $\mathbb{R}$. This isomorphism is defined by

$$j(v \otimes w)(k, l) := \langle k, v \rangle \langle l, w \rangle. \tag{3.4.2}$$

This can be regarded as an extension of the well-known fact mentioned earlier (see Eq. (3.3.5)) that, if $V$ is finite-dimensional, there is an isomorphism $\chi : V \to (V^*)^*$ given by

$$\begin{aligned} \chi : V &\to V^{**} \\ \chi(v)(k) &:= \langle k, v \rangle \end{aligned} \tag{3.4.3}$$

[4]Most of the material that follows is only well-defined for vector spaces whose dimension is finite. The tensor product of infinite-dimensional topological vector spaces requires careful consideration of the topologies concerned.

for all vectors $v \in V$ and $k \in V^*$.

Applying these ideas in the context of differential geometry leads to the following definition.

**Definition 3.12**

A *tensor of type* $(r, s)$ at a point $p \in \mathcal{M}$ is an element of the tensor product space

$$T_p^{r,s}\mathcal{M} := \left[\overset{r}{\otimes}\, T_p\mathcal{M}\right] \otimes \left[\overset{s}{\otimes}\, T_p^*\mathcal{M}\right]. \tag{3.4.4}$$

**Comments**

1. The notation $\overset{r}{\otimes}\, V$ means the vector space formed by taking the tensor product of $V$ with itself $r$-times. It is convenient to allow this to include the value $r = 0$, with the product in this case being defined to be $\mathbb{R}$. In using this convention it should be noted that there is a canonical isomorphism $W \otimes \mathbb{R} \cong W$ for any real vector space $W$.

2. Using Eq. (3.4.2) and Eq. (3.4.3) for the finite-dimensional spaces $T_p\mathcal{M}$ and $T_p^*\mathcal{M}$ it follows that an alternative definition of a tensor of type $(r, s)$ at the point $p \in \mathcal{M}$ is as a multilinear map

$$\overset{r}{\times}\, T_p^*\mathcal{M} \;\times\; \overset{s}{\times}\, T_p\mathcal{M} \longrightarrow \mathbb{R} \tag{3.4.5}$$

from the Cartesian product of $T_p^*\mathcal{M}$ $r$-times, and $T_p\mathcal{M}$ $s$-times, into the real numbers.

3. Special cases of type $(r, s)$ tensors include

(i) $T_p^{0,1}\mathcal{M} = T_p^*\mathcal{M}$

(ii) $T_p^{1,0}\mathcal{M} = (T_p^*\mathcal{M})^* \cong T_p\mathcal{M}$

(iii) $T_p^{r,0}\mathcal{M}$ is called the space of $r$-*contravariant* tensors

(iv) $T_p^{0,s}\mathcal{M}$ is called the space of $s$-*covariant* tensors.

4. The definition extends at once to include the idea of a $(r, s)$-*tensor field* on $\mathcal{M}$. This is simply a 'smooth' assignment of a tensor of type $(r, s)$ to each point $p \in \mathcal{M}$. (Exercise: What would 'smooth' mean in the case when (i) the tensor product is defined by Eq. (3.4.4); and (ii) when Eq. (3.4.5) is used instead?)

5. In the abstract definition (3.4.1) of the tensor product $V \otimes W$ of two vector spaces $V$ and $W$, the tensor product of $v \in V$ and $w \in W$ (denoted $v \otimes w$) is defined to be the element $\mu(v, w) \in V \otimes W$.

In the more concrete definition (3.4.2), the tensor product of $v$ and $w$ is defined to be the unique bilinear map from $V^* \times W^*$ into $\mathbb{R}$ that takes on the value $\langle k, v\rangle\langle l, w\rangle$ on $(k, l) \in V^* \times W^*$.

This definition extends to multiple tensor products. In particular, if $\alpha \in T_p^{r,s}\mathcal{M}$ and $\alpha' \in T_p^{r',s'}\mathcal{M}$ then $\alpha \otimes \alpha' \in T_p^{r+r',s+s'}\mathcal{M}$ is defined to be the multilinear map

$$\begin{aligned} &\alpha \otimes \alpha'(k_1, \ldots, k_{r+r'}; v_1, \ldots, v_{s+s'}) := \\ &\qquad \alpha(k_1, \ldots, k_r; v_1, \ldots, v_s)\, \alpha'(k_{r+1}, \ldots, k_{r+r'}; v_{s+1}, \ldots, v_{s+s'}). \end{aligned} \tag{3.4.6}$$

6. If $\dim V < \infty$ and $\dim W < \infty$ then $\dim(V \otimes W) = \dim V \dim W$. This is clear from the fact that if $\{e_1, e_2, \ldots, e_n\}$ is a basis set for $V$ and if $\{f_1, f_2, \ldots, f_m\}$ is a basis set for $W$ (so $n = \dim V$ and $m = \dim W$) then a basis set for $V \otimes W$ is $\{e_i \otimes f_j \mid i = 1, \ldots, n \text{ and } j = 1, \ldots, m\}$.

In particular, a basis set for $T_p^{r,s}\mathcal{M}$ with respect to a local coordinate system is the set of all vectors of the form

$$\left(\frac{\partial}{\partial x^{\mu_1}}\right)_p \otimes \cdots \otimes \left(\frac{\partial}{\partial x^{\mu_r}}\right)_p \otimes (dx^{\nu_1})_p \otimes \cdots \otimes (dx^{\nu_s})_p, \tag{3.4.7}$$

and the expansion coefficients $\alpha^{\mu_1 \ldots \mu_r}_{\nu_1 \ldots \nu_s}$ of $\alpha \in T_p^{r,s}\mathcal{M}$ with respect to this basis are

$$\alpha^{\mu_1 \ldots \mu_r}_{\nu_1 \ldots \nu_s} = \alpha\left((dx^{\mu_1})_p, \ldots, (dx^{\mu_r})_p\,; \left(\frac{\partial}{\partial x^{\nu_1}}\right)_p, \ldots, \left(\frac{\partial}{\partial x^{\nu_s}}\right)_p\right) \tag{3.4.8}$$

in which $\alpha \in T_p^{r,s}\mathcal{M}$ is viewed as a multilinear map from the product of $T_p^*\mathcal{M}$ $r$-times and $T_p\mathcal{M}$ $s$-times. □

### 3.4.2 The idea of an $n$-form

A very important example of a tensor field is an '$n$-form' where $0 \leq n \leq \dim \mathcal{M}$. A 0-form is defined to be a function in $C^\infty(\mathcal{M})$, and a 1-form was defined in Section 3.3. The formal definition of a general $n$-form is as follows.

**Definition 3.13**

An *$n$-form* is a tensor field $\omega$ of type $(0, n)$ that is totally skew-symmetric in the sense that, for any permutation $P$ of the indices $1, 2, \ldots, n$,

$$\omega(X_1, X_2, \ldots, X_n) = (-1)^{\deg(P)} \omega(X_{P(1)}, X_{P(2)}, \ldots, X_{P(n)}) \tag{3.4.9}$$

where $X_1, X_2, \ldots, X_n$ are arbitrary vector fields on $\mathcal{M}$ and $\deg(P)$ is the degree of the permutation $P$, *i.e.*, it is $+1$ if $P$ is even and $-1$ if $P$ is odd.

We shall write the set of all $n$-forms on $\mathcal{M}$ as $A^n(\mathcal{M})$.

We shall be interested in applying the tensor-product operation to $n$-forms. However, if $\omega_1 \in A^{n_1}(\mathcal{M})$ and $\omega_2 \in A^{n_2}(\mathcal{M})$ then $\omega_1 \otimes \omega_2$ is an $(n_1+n_2)$-covariant tensor field but it will not be an $(n_1+n_2)$-form since it does not satisfy the alternating property with respect to all its 'indices'; *i.e.*, $\omega_1 \otimes \omega_2(X_1, \ldots, X_{n_1}; X_{n_1+1}, \ldots, X_{n_1+n_2})$ will not be antisymmetric under permutations of the indices that take an index $k$ in the range $1 \leq k \leq n_1$ into the range $n_1 + 1 \leq k \leq n_1 + n_2$. This is remedied with the aid of the following definition:

**Definition 3.14**

If $\omega_1 \in A^{n_1}(\mathcal{M})$ and $\omega_2 \in A^{n_2}(\mathcal{M})$, the *wedge product*, or *exterior product*, of $\omega_1$ and $\omega_2$ is the $(n_1 + n_2)$-form $\omega_1 \wedge \omega_2$ defined by

$$\omega_1 \wedge \omega_2 := \frac{1}{n_1! n_2!} \sum_{\text{Perms } P} (-1)^{\deg(P)} (\omega_1 \otimes \omega_2)^P \tag{3.4.10}$$

where, if $\omega$ is any tensor field of type $(0, n)$, the 'permuted' tensor field $\omega^P$ of type $(0, n)$ is defined to be

$$\omega^P(X_1, X_2, \ldots, X_n) := \omega(X_{P(1)}, X_{P(2)}, \ldots, X_{P(n)}) \qquad (3.4.11)$$

for all vector fields $X_1, X_2, \ldots, X_n$ on the manifold $\mathcal{M}$.

## Comments

1. This wedge product has many important properties with respect to the other structures that can be imparted to differential forms. For example, the pull-back of a 1-form defined in Eq. (3.3.18) can be generalised in an obvious way to an arbitrary $n$-form. (In the case of a function/0-form, the pull-back of $f \in C^\infty(\mathcal{N})$ by $h : \mathcal{M} \to \mathcal{N}$ is defined to be the function $f \circ h$ in $C^\infty(\mathcal{M})$. *cf.* Eq. (3.3.37) in this context.)

This generalised pull-back has the very significant property of 'commuting' with the wedge product. More precisely, if $h : \mathcal{M} \to \mathcal{N}$ and if $\alpha$ and $\beta$ are differential forms on $\mathcal{N}$, then

$$h^*(\alpha \wedge \beta) = (h^*\alpha) \wedge (h^*\beta). \qquad (3.4.12)$$

Another major example of structure preserved by pull-back will appear below in the discussion of the exterior derivative of a general $n$-form.

2. The wedge product turns the vector space $A(\mathcal{M}) := \oplus_{n=0}^{\dim \mathcal{M}} A^n(\mathcal{M})$ into a graded algebra with

$$\omega_1 \wedge \omega_2 = (-1)^{n_1 n_2} \omega_2 \wedge \omega_1 \qquad (3.4.13)$$

if $\omega_1 \in A^{n_1}(\mathcal{M})$ and $\omega_2 \in A^{n_2}(\mathcal{M})$.

3. It can be shown [Exercise!] that a basis set for the $n$-forms evaluated at a point $p \in \mathcal{M}$ is the set of all vectors of the form

$$(dx^{\mu_1})_p \wedge (dx^{\mu_2})_p \wedge \cdots \wedge (dx^{\mu_n})_p. \qquad (3.4.14)$$

Furthermore, if a differential form $\omega$ expanded locally in terms of such elements then the coefficients in the expansion $\omega_{\mu_1 \mu_2 \cdots \mu_n}$ are skew-symmetric under the exchange of any pair of indices. □

### 3.4.3 The definition of the exterior derivative

In Eq. (3.3.35) we defined the exterior derivative $df$ of a 0-form $f \in C^\infty(\mathcal{M})$ as a 1-form and showed that it commuted with pull-backs in the sense of Eq. (3.3.37). We wish now to extend this definition to an arbitrary $n$-form and to construct an 'exterior derivative' that will convert it into an $(n+1)$-form in a way that commutes with pull-backs.

**Definition 3.15**

If $\omega$ is an $n$-form on $\mathcal{M}$ with $1 \leq n < \dim \mathcal{M}$ then the *exterior derivative* of $\omega$ is the $(n+1)$-form $d\omega$ defined by

$$\begin{aligned} d\omega(X_1, \ldots, X_{n+1}) &:= \sum_{i=1}^{n+1} (-1)^{i+1} X_i(\omega(X_1, \ldots, \not{X}_i \ldots, X_{n+1})) \\ &+ \sum_{i<j} (-1)^{i+j} \omega([X_i, X_j], X_1, \ldots, \not{X}_i, \ldots, \not{X}_j, \ldots, X_{n+1}) \end{aligned}$$

for all vector fields $X_1, X_2, \ldots, X_{n+1}$. The symbol $\not{X}$ means that the particular field $X$ in the series is omitted.

**Comments**

1. There are no $n$-forms for $n > \dim\mathcal{M}$ and therefore it is conventional to define $d\omega := 0$ if $\omega$ is a $\dim\mathcal{M}$-form.

2. The case of a 1-form is of particular importance in the theory of fibre bundles and Lie groups. The complicated formula in Eq. (3.4.15) then simplifies considerably, and the 2-form $d\omega$ acting on any pair of vector fields $X, Y$ is

$$d\omega(X, Y) = X(\langle\omega, Y\rangle) - Y(\langle\omega, X\rangle) - \langle\omega, [X, Y]\rangle. \tag{3.4.15}$$

Note that the notation $X(\langle\omega, Y\rangle)$ means the effect of acting with the vector field $X$ on the function $\langle\omega, Y\rangle$ in $C^\infty(\mathcal{M})$.

3. If an $n$-form $\omega$ is expanded locally as $\omega = \omega_{\mu_1\mu_2,\ldots,\mu_n} dx^{\mu_1} \wedge dx^{\mu_2} \wedge \cdots \wedge dx^{\mu_n}$ (with all indices summed over) then

$$d\omega = \omega_{\mu_1\mu_2,\ldots,\mu_n\,,\nu}\, dx^\nu \wedge dx^{\mu_1} \wedge dx^{\mu_2} \wedge \cdots \wedge dx^{\mu_n}. \tag{3.4.16}$$

4. The exterior derivative behaves rather nicely with respect to the wedge product and the associated graded algebra structure on $A(\mathcal{M})$; specifically

$$d(\omega_1 \wedge \omega_2) = d\omega_1 \wedge \omega_2 + (-1)^{\deg \omega_1} \omega_1 \wedge d\omega_2. \tag{3.4.17}$$

5. The definition (3.4.15) of exterior differentiation gives commutativity with the act of pulling back: if $h : \mathcal{M} \to \mathcal{N}$ and $\omega$ is an $n$-form on $N$ then

$$d(h^*\omega) = h^*(d\omega). \tag{3.4.18}$$

□

### 3.4.4 The local nature of the exterior derivative

An important, but not obvious, property of the exterior derivative is that the value at a point $p \in \mathcal{M}$ of the function $d\omega(X_1, X_2, \ldots, X_{n+1})$ defined in Eq. (3.4.15), depends only on the values of the vector fields $X_1, X_2, \ldots X_n$ at the point $p$. This is a corollary to the result that, for any $f \in C^\infty(\mathcal{M})$, the exterior derivative is '$f$-linear' in the sense that

$$d\omega(X_1, X_2, \ldots, fX_i, \ldots X_{n+1}) = f d\omega(X_1, X_2, \ldots, X_i, \ldots X_{n+1}) \tag{3.4.19}$$

for all $i = 1, 2, \ldots, n+1$. For the sake of simplicity we shall only prove this result for the case of the exterior derivative of a one-form.

**Theorem 3.2** *If $\omega$ is a one-form on a manifold $\mathcal{M}$, the exterior derivative $d\omega$ is $C^\infty(\mathcal{M})$-linear in the sense that, for all $f \in C^\infty(\mathcal{M})$,*

$$d\omega(fX, Y) = f d\omega(X, Y) \tag{3.4.20}$$

*for all vector fields $X$ and $Y$ on $\mathcal{M}$.*

**Proof**

Using Eq. (3.4.15), we have

$$d\omega(fX, Y) = fX(\langle \omega, Y \rangle) - Y(\langle \omega, fX \rangle) - \langle \omega, [fX, Y] \rangle, \tag{3.4.21}$$

and the first observation is that $(fX)(h) = fX(h)$ for all $h \in C^\infty(\mathcal{M})$, *i.e.*, $(fX)_p = f(p)X_p$. Secondly, we note that

$$\langle \omega, fX \rangle_p = \langle \omega_p, (fX)_p \rangle = \langle \omega_p, f(p)X_p \rangle = f(p)\langle \omega_p, X_p \rangle \qquad (3.4.22)$$

where the last step follows because $\omega_p$ is $\mathbb{R}$-linear. Therefore, as functions on $\mathcal{M}$,

$$\langle \omega, fX \rangle = f\langle \omega, X \rangle \qquad (3.4.23)$$

and so

$$Y(\langle \omega, fX \rangle) = Y(f)\langle \omega, X \rangle + fY(\langle \omega, X \rangle) \qquad (3.4.24)$$

using the derivation property of the vector field $Y$.

We also have, for all $g \in C^\infty(\mathcal{M})$,

$$\begin{aligned} [fX, Y](g) &= fX(Y(g)) - Y(fX(g)) \\ &= fX(Y(g)) - Y(f)X(g) - fY(X(g)) \end{aligned} \qquad (3.4.25)$$

using the derivation property once more. Thus $[fX, Y] = f[X, Y] - Y(f)X$, and so, using Eq. (3.4.23), we get

$$\langle \omega, [fX, Y] \rangle = f\langle \omega, [X, Y] \rangle - Y(f)\langle \omega, X \rangle. \qquad (3.4.26)$$

However, when Eq. (3.4.24) and Eq. (3.4.26) are inserted in Eq. (3.4.21), the $Y(f)\langle \omega, X \rangle$ terms cancel, and we end up with the relation $d\omega(fX, Y) = f d\omega(X, Y)$. **QED**

**Comments**

1. The equality $d\omega(X, Y) = -d\omega(Y, X)$ means that

$$d\omega(X, fY) = -d\omega(fY, X) = -f d\omega(Y, X) = f d\omega(X, Y), \qquad (3.4.27)$$

so that $d\omega(X, Y)$ is $f$-linear in $Y$ as well as in $X$.

2. To show that $d\omega(X, Y)(p)$ depends only on the values of $X$ and $Y$ at the point $p \in \mathcal{M}$, choose a coordinate chart $(U, \phi)$ in a neighbourhood of $p \in \mathcal{M}$. Then on the open set $U$, the vector field at $q \in U$ can be written as $X_q = \sum_{\mu=1}^m X^\mu(q)(\partial_\mu)_q$, and so $d\omega(X, Y)(p) =$

$d\omega(\sum_{\mu=1}^m X^\mu \partial_\mu, \sum_{\nu=1}^m Y^\nu \partial_\nu)(p) = \sum_{\mu=1}^m \sum_{\nu=1}^m d\omega(X^\mu \partial_\mu, Y^\nu \partial_\nu)(p)$. But, according to the theorem, for each fixed value for the indices $\mu$ and $\nu$, we have

$$d\omega(X^\mu \partial_\mu, Y^\nu \partial_\nu)(p) = X^\mu(p) Y^\nu(p) d\omega(\partial_\mu, \partial_\nu)(p) \qquad (3.4.28)$$

and so $d\omega(X, Y)(p)$ depends only on the values $X^\mu(p)$, $Y^\nu(p)$ of the components of the vector fields $X, Y$ at $p \in \mathcal{M}$. But $X^\mu(p) = X_p(x^\mu)$, and so these components (and hence $d\omega_p$) depend only on the values of the vector fields at $p$. **QED**

□

## 3.5 DeRham Cohomology

One of the most important mathematical applications of differential forms stems from the observation that if the exterior operator $d$ is applied twice to any $n$-form $\omega$ then the resulting $(n+2)$-form $d(d(\omega))$ is identically zero. This result is usually written succinctly as

$$d^2 = 0 \qquad (3.5.1)$$

and is of the greatest significance.[5] It can be proved directly from the general formula (3.4.15) for the exterior derivative; in local coordinates it is a simple consequence of the fact that mixed partial derivatives commute.

A central ingredient in DeRham theory is the sequence of real vector spaces

$$0 \xrightarrow{d} C^\infty(\mathcal{M}) \xrightarrow{d} A^1(\mathcal{M}) \xrightarrow{d} A^2(\mathcal{M}) \xrightarrow{d} \dots \xrightarrow{d} A^m(\mathcal{M}) \xrightarrow{d} 0 \qquad (3.5.2)$$

in which the product of any two consecutive maps is zero by virtue of Eq. (3.5.1). Note that, for convenience, I have effectively introduced a space $A^{-1}(\mathcal{M}) := 0$ with $d : A^{-1}(\mathcal{M}) \to A^0(\mathcal{M}) = C^\infty(\mathcal{M})$ being defined to be the trivial map.

[5] A special case of Eq. (3.5.1) is the well-known identity $\text{curl}(\text{grad}(\phi)) = 0$ in standard vector calculus.

This sequence of maps and spaces is called the *DeRham complex* of the differentiable manifold $\mathcal{M}$. It is a special example of the more general concept of a *differential complex* which is defined to be any sequence of maps and vector spaces (unconnected with differential geometry in general) of the form

$$\ldots \longrightarrow A \xrightarrow{a} B \xrightarrow{b} C \xrightarrow{c} D \xrightarrow{d} E \xrightarrow{e} F \xrightarrow{f} \ldots \tag{3.5.3}$$

with the property that the composition of any consecutive pair of the linear maps is zero. Thus

$$b \circ a = c \circ b = d \circ c = e \circ d = f \circ e = \ldots = 0 \tag{3.5.4}$$

or, equivalently,

$$\mathrm{Im}(a) \subset \mathrm{Ker}(b),\ \mathrm{Im}(b) \subset \mathrm{Ker}(c),\ \mathrm{Im}(c) \subset \mathrm{Ker}(d), \ldots \tag{3.5.5}$$

as shown in Figure 3.7.

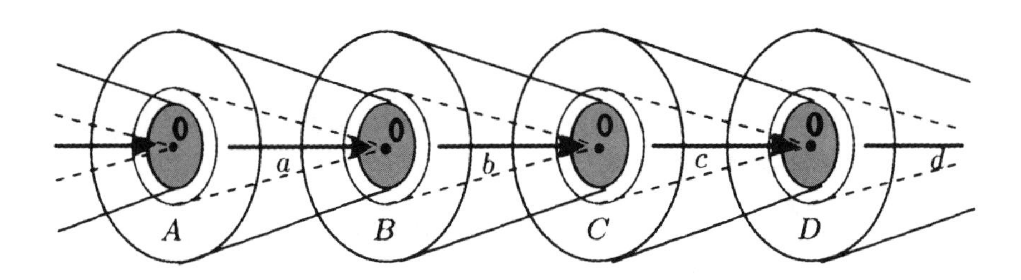


Figure 3.7: A differential complex.

The sequence of inclusions in Eq. (3.5.5) has a direct parallel in the case of differential forms. Specifically, the property $d^2 = 0$ implies that in the sequence (3.5.2) there are the subspace inclusions

$$\mathrm{Im}(d : A^{n-1}(\mathcal{M}) \to A^n(\mathcal{M})) \subset \mathrm{Ker}(d : A^n(\mathcal{M}) \to A^{n+1}(\mathcal{M})). \tag{3.5.6}$$

In general, whenever relations of this type arise there is always an associated cohomology theory in which the cohomology groups measure the extent to which the image set in Eq. (3.5.6) is a proper subset of the kernel set. In the present context of differential forms this idea is captured in the following central definitions of the DeRham theory.[6]

**Definition 3.16**

1. An $n$-form $\omega$ is *closed* if $d\omega = 0$. The set of all closed $n$-forms is denoted $Z^n(\mathcal{M})$; thus

$$\begin{aligned} Z^n(\mathcal{M}) &:= \operatorname{Ker}\left(d : A^n(\mathcal{M}) \to A^{n+1}(\mathcal{M})\right) \\ &= \{\omega \in A^n(\mathcal{M}) \mid d\omega = 0\}. \end{aligned} \tag{3.5.7}$$

   A closed $n$-form is also said to be an *n-cocycle* for the DeRham cohomology.

2. An $n$-form $\omega$ is *exact* if there exists some $(n-1)$-form $\beta$ such that $d\beta = \omega$. The set of exact $n$-forms is denoted $B^n(\mathcal{M})$; thus

$$\begin{aligned} B^n(\mathcal{M}) &= \operatorname{Im}\left(d : A^{n-1}(\mathcal{M}) \to A^n(\mathcal{M})\right) \\ &= \{\omega \in A^n(\mathcal{M}) \mid \omega = d\beta \text{ for some } \beta \in A^{n-1}(\mathcal{M})\}. \end{aligned} \tag{3.5.8}$$

   An exact $n$-form is also said to be an *n-coboundary* for the DeRham cohomology.

It is clear from Eq. (3.5.6) that every exact $n$-form is closed, and hence the vector space $B^n(\mathcal{M})$ is a subspace of $Z^n(\mathcal{M})$. But when are they equal? This question of when a closed form is also exact has far reaching implications since the problem of proving that a given closed $n$-form is exact involves global topological properties of the manifold. Indeed, it can be shown (Poincaré's lemma) that on a Euclidean space $\mathbb{R}^m$ every closed $n$-form is exact for $n > 0$. Hence the failure of $Z^n(\mathcal{M})$ to equal $B^n(\mathcal{M})$ does indeed reflect the degree to which the manifold $\mathcal{M}$ is not topologically trivial in the sense that the space $\mathbb{R}^m$ is (specifically, $\mathbb{R}^m$ is contractible). For this reason it is desirable to find some precise mathematical construction that faithfully reflects the departure from exactness, and this is contained in the following definition.

---

[6]For a comprehensive discussion see Bott & Tu (1982)

**Definition 3.17**

The *DeRham cohomology groups* $H^n_{DR}(\mathcal{M})$, $0 \leq n \leq \dim(\mathcal{M})$, of a manifold $\mathcal{M}$ are the quotient vector spaces

$$H^n_{DR}(\mathcal{M}) := Z^n(\mathcal{M})/B^n(\mathcal{M}). \tag{3.5.9}$$

**Comments**

1. Since $A^{-1}(\mathcal{M}) := \{0\}$ it follows that $B^0(\mathcal{M}) = \{0\}$. On the other hand, a function $f \in A(\mathcal{M})$ is closed if and only if its partial derivatives vanish in any coordinate system. Since $\mathcal{M}$ is connected (by assumption) it follows that a closed 0-form is simply some constant function, *i.e.*, a real number. Thus

$$B^0(\mathcal{M}) \cong \{0\} \text{ and } Z^0(\mathcal{M}) \cong \mathbb{R} \tag{3.5.10}$$

and hence

$$H^0_{DR}(\mathcal{M}) \cong \mathbb{R} \tag{3.5.11}$$

for all connected manifolds $\mathcal{M}$.

If $\mathcal{M}$ is not connected but has $k$ connected components, then it is straightforward to see that Eq. (3.5.11) generalises to

$$H^0_{DR}(\mathcal{M}) \cong \mathbb{R}^k \tag{3.5.12}$$

which can be thought of as the direct sum of $k$ copies of the real line $\mathbb{R}$.

2. On $\mathcal{M} = \mathbb{R}$, the most general 1-form can be written as $\omega_x = g(x)(dx)_x$ with respect to the global coordinate system on $\mathbb{R}$. But then

$$d\omega = \frac{\partial}{\partial x} g(x) dx \wedge dx + g(x) d(dx) = 0 \tag{3.5.13}$$

and so every 1-form is closed (as expected, since $\dim \mathbb{R} = 1$).

However, if we define

$$f(x) := \int_0^x g(y)\, dy \tag{3.5.14}$$

then $df = g(x)dx$ and hence the 1-form $\omega$ is necessarily exact. This implies that $H^1_{DR}(\mathbb{R}) \cong \{0\}$, and indeed, since the Poincaré lemma implies that for any manifold $\mathbb{R}^m$ we have $Z^n(\mathbb{R}^m) \cong B^n(\mathbb{R}^m)$ for all $n > 0$, it follows that

$$\begin{aligned} H^n_{DR}(\mathbb{R}^m) &\cong \{0\} \quad \text{for } n > 0 \\ &\cong \mathbb{R} \quad\ \ \text{for } n = 0. \end{aligned} \tag{3.5.15}$$

3. Now consider the one-dimensional manifold $S^1$. Since this is connected we have $H^0_{DR}(S^1) \cong \mathbb{R}$, as before. And, since $S^1$ is one-dimensional, it is also true that every 1-form is necessarily closed, so that $A^1(S^1) \cong Z^1(S^1)$.

However, unlike the case $\mathcal{M} = \mathbb{R}$, it is not true that every 1-form is exact. For example, in angular coordinates, the one-form $d\theta$ is not exact (not withstanding its appearance) since the angular variable $\theta$ is not continuous or differentiable when considered as a function on the entire space $S^1$, and there is no differentiable function on the circle such that $d\theta = df$.[7] It is not difficult to deduce from this that

$$H^1_{DR}(S^1) \cong \mathbb{R}, \tag{3.5.16}$$

which reflects the most important topological difference between the circle and the real line: namely, the former is not simply connected but has $\pi_1(S^1) \cong \mathbf{Z}$.

4. In Section 3.4 it was shown that exterior differentiation commutes with the act of pulling-back in the sense that (Eq. (3.4.18))

$$d(h^*\omega) = h^*(d\omega) \tag{3.5.17}$$

where $h : \mathcal{M} \to \mathcal{N}$ and $\omega$ is an $n$-form on $\mathcal{N}$. This equation has a very significant implication for the cohomological use of differential forms since it implies that

(i) if $\omega$ is closed then so is $h^*\omega$;

(ii) if $\omega$ is an exact form with $\omega = d\beta$ then $h^*(d\beta) = d(h^*\beta)$ and hence $h^*\omega$ is also exact.

[7]Examples of functions on the circle that *are* globally-defined and differentiable, are $\sin n\theta$ and $\cos n\theta$.

Now, suppose that $\omega_1$ and $\omega_2$ are $n$-forms that belong to the same equivalence class in the quotient space $Z^n(\mathcal{N})/B^n(\mathcal{N})$. Then there must be some $(n-1)$-form $\beta$ such that $\omega_1 - \omega_2 = d\beta$. But then the two results above imply that the pull-backs $h^*(\omega_1)$ and $h^*(\omega_2)$ lie in the same equivalence class in $Z^n(\mathcal{M})/B^n(\mathcal{M})$. In summary, a differentiable function $h : \mathcal{M} \to \mathcal{N}$ induces a linear map

$$h^* : H^n_{DR}(\mathcal{N}) \to H^n_{DR}(\mathcal{M}). \tag{3.5.18}$$

A procedure of this type is an intrinsic feature of any useful cohomology theory: specifically, the cohomology groups/vector spaces probe the topological properties of the manifold or topological space with which they are associated, and the properties of maps between such spaces are reflected in the induced homomorphisms between the groups attached to the spaces.

5. Two continuous functions $f$ and $g$ between topological spaces $X$ and $Y$ are said to be *homotopic* if they can be 'deformed' into each other continuously. More precisely, there must be some continuous function $F : [0,1] \times X \to Y$ such that $F(0,x) = f(x)$ and $F(1,x) = g(x)$.

A characteristic property of traditional cohomology theories (using simplicial decompositions, for example) is that under these circumstances the associated homomorphisms

$$f^* : H^n(Y) \to H^n(X) \text{ and } g^* : H^n(Y) \to H^n(X) \tag{3.5.19}$$

are in fact equal. The important result for us is that this property can be shown to hold also for the DeRham cohomology of a differentiable manifold. Thus, if $f$ and $g$ are two differentiable functions from a manifold $\mathcal{M}$ to a manifold $\mathcal{N}$ then they induce identical homomorphisms (3.5.18) if they are homotopic to each other.

6. It follows from the above that the DeRham cohomology theory for differentiable manifolds has many general properties that are shared by the usual simplicial, or singular, cohomology structures. This is no coincidence. Indeed, one of the most remarkable and powerful results in the theory of the global structure of differential manifolds is DeRham's theorem which asserts that the DeRham cohomology

groups are actually isomorphic to the real cohomology groups computed using singular or simplicial theory. This really is very striking since the singular/simplicial groups depend only on the underlying topological structure of a manifold (*i.e.*, they know nothing of its differential structure of coordinate charts) while, on the other hand, differential forms are basically defined in a local way that rests heavily on the differential structure: the ability to probe global properties stems entirely from the special properties of the operation of exterior differentiation.

This identification of the DeRham and 'topological' cohomology groups can be used either way. The question of the existence of closed but non-exact $n$-forms can be answered at once if the (say) real, simplicial group $H^n(\mathcal{M}, \mathbb{R})$ is known. Alternatively, there may be cases where the simplicial groups can be computed using the differential forms on the manifold. From all perspectives, DeRham's theorem is of major importance.

7. Another satisfying link between the various cohomology theories concerns the 'cup product'. In the conventional singular/simplicial theory this is a pairing homomorphism

$$\cup : H^p(X) \otimes H^q(X) \to H^{p+q}(X) \tag{3.5.20}$$

and it is the existence of this entity that makes cohomology potentially more powerful than homology.

Is there an analogue of this cup product for the DeRham theory? A clue is given by the skew commutativity of the product in the simplicial/singular case:

$$a \cup b = (-1)^{pq} b \cup a \tag{3.5.21}$$

if $a \in H^p(X)$ and $b \in H^q(X)$. This is strikingly reminiscent of Eq. (3.4.13) which states that differential forms have a similar graded structure under the wedge product. What is suggested therefore is that, in DeRham cohomology, the wedge product plays the role of the cup product.

This possibility is strongly supported by Eq. (3.4.17) which means that the conditions of closedness and exactness are preserved under the wedge product in such a way as to induce a linear map

$$\wedge : H^p_{DR}(\mathcal{M}) \otimes H^q_{DR}(\mathcal{M}) \to H^{p+q}_{DR}(\mathcal{M}) \tag{3.5.22}$$

that satisfies Eq. (3.5.21). The full version of DeRham's theorem confirms this conjecture and shows that the wedge product is indeed an exact analogue for differential form cohomology of the cup product in the other theories.

# Chapter 4

# Lie Groups

## 4.1 The Basic Ideas

### 4.1.1 The first definitions

The key idea of a Lie group is that it is a group in the usual sense, but with the additional property that it is also a differentiable manifold, and in such a way that the group operations are 'smooth' with respect to this structure. A good example is the set $\{z \in \mathbb{C} \mid |z|^2 = 1\}$ of all complex numbers of modulus one. This is clearly a group under the action of multiplication, but it is also a manifold: namely, the circle $S^1$.

Lie groups are of the greatest importance in modern theoretical physics, and the subject is one that can be approached from a variety of perspectives. The discussion that follows is relatively elementary and is concerned mainly with (i) the geometrical relation between a Lie group and its Lie algebra; and (ii) those aspects of Lie group theory that are especially appropriate for the section of the course that deals with fibre bundle theory. It should really be supplemented by private reading on the general theory of Lie algebras, and their use in modern theoretical physics (for example, in the classification of elementary particles).

The formal definition of a Lie group is as follows.

**Definition 4.1**

1. A *real Lie group* $G$ is a set that is

   (a) a group in the usual algebraic sense;

   (b) a differentiable manifold with the properties that taking the product of two group elements, and taking the inverse of a group element, are smooth operations. Specifically, the maps

$$\begin{aligned} \mu : G \times G &\rightarrow G \\ (g_1, g_2) &\mapsto g_1 g_2 \end{aligned} \tag{4.1.1}$$

   and

$$\begin{aligned} \iota : G &\rightarrow G \\ g &\mapsto g^{-1} \end{aligned} \tag{4.1.2}$$

   are both $C^\infty$.

2. A *Lie subgroup* of $G$ is a subset $H$ of $G$ that is (i) a subgroup of the group $G$ in the algebraic sense; (ii) a submanifold of the differentiable manifold $G$; and (iii) a topological group with respect to the subspace topology.

3. The *right* and *left translations* of $G$ are diffeomorphisms[1] of $G$ labelled by the elements $g \in G$ and defined by

$$\begin{aligned} r_g : G &\rightarrow G, & l_g : G &\rightarrow G \\ g' &\mapsto g'g & g' &\mapsto gg' \end{aligned} \tag{4.1.3}$$

**Comments**

1. The maps $\mu : G \times G \rightarrow G$ and $\iota : G \rightarrow G$ can be used to express the basic axioms of a group purely in map-theoretic terms.

[1]Note that left and right translations exist for any group $G$, not just for Lie groups. In the algebraic case they are only required to be bijections; if $G$ is a topological group, they are required to be homeomorphisms.

For example, the associativity condition $g_1(g_2g_3) = (g_1g_2)g_3$ for all $g_1, g_2, g_3 \in G$, is equivalent to the relation

$$\mu \circ (\mathrm{id}_G \times \mu) = \mu \circ (\mu \times \mathrm{id}_G) \tag{4.1.4}$$

between the maps $G \times G \times G \xrightarrow{\mathrm{id}_G \times \mu} G \times G \xrightarrow{\mu} G$ and $G \times G \times G \xrightarrow{\mu \times \mathrm{id}_G} G \times G \xrightarrow{\mu} G$, where $\mathrm{id}_G : G \to G$ denotes the identity map on $G$. [Exercise: translate the remaining axioms for a group into this language.]

Writing the axioms in this way is sometimes technically convenient in standard group theory. It also allows a generalisation to a 'group-like object' in a category which is other than the category of sets.

2. A somewhat weaker notion is that of a *topological group*. This is defined to be a group $G$ that is at the same time a topological space, and in such a way that the maps $\mu : G \times G \to G$ and $\iota : G \to G$ are continuous. Every Lie group is a topological group, but the converse is not true.

3. There is an analogous theory of *real analytic* Lie groups and *complex analytic* Lie groups in which the underlying manifold is required to be a real or complex analytic manifold respectively.

It is a very deep and non-trivial theorem that every real $C^\infty$ Lie group admits a real analytic structure in the sense that a set of coordinate charts covering $G$ can always be found with the property that the associated overlap functions are $C^\omega$. This famous result is due to Montgomery, Gleason and Zippin (Montgomery & Zippin 1955); it constitutes the solution to Hilbert's fifth problem.

4. The left and right translations satisfy the relations

$$l_{g_1} \circ l_{g_2} = l_{g_1 g_2} \tag{4.1.5}$$

and

$$r_{g_1} \circ r_{g_2} = r_{g_2 g_1} \tag{4.1.6}$$

respectively. Furthermore, the maps $g \mapsto l_g$ and $g \mapsto r_g$ are both injective and hence define an isomorphism and an anti-isomorphism[2] respectively from $G$ into the group Diff$(G)$ of diffeomorphisms of $G$.

In general, a *homomorphism* between two Lie groups $G_1$ and $G_2$ (*i.e.*, a structure-preserving map) is defined to be a group homomorphism that is also a smooth map between the two underlying differentiable manifolds. This also applies to the maps $G \mapsto$ Diff$(G)$ afforded by the left and right translations in so far as the infinite-dimensional group Diff$(G)$ can be equipped with the differential structure of a genuine Lie group. □

**Examples**

1. When equipped with its natural differential structure, the vector space $\mathbb{R}^n$ is an $n$-dimensional (abelian) Lie group.

2. The group of *Möbius transformations* of the complex plane $\mathbb{C}$ is defined to be the subset of diffeomorphisms with the special form

$$z \mapsto \frac{az+b}{cz+d} \quad \text{where } a,b,c,d \in \mathbb{C} \text{ with } ad-bc=1. \tag{4.1.7}$$

This 6-dimensional real Lie group plays an important role in the geometrical approach to complex analysis and in the theory of relativistic strings.

3. The circle $S^1 := \{z \in \mathbb{C} \mid |z| = 1\}$ can be given the structure of a real one-dimensional Lie group by employing the usual differential structure, and defining the group combination law as multiplication of complex numbers. This group is usually denoted $U(1)$ and belongs to a series $U(n)$, $n = 1, 2, \ldots$ of Lie groups (see below). As a topological space, it is compact.

4. The *real, general linear group* in $n$-dimensions is defined as

$$GL(n,\mathbb{R}) := \{A \in M(n,\mathbb{R}) \mid \det A \neq 0\} \tag{4.1.8}$$

[2] A map $\phi : G_1 \to G_2$ between groups $G_1$ and $G_2$ is said to be an *anti-homomorphism* if $\phi(ab) = \phi(b)\phi(a)$ for all $a, b \in G_1$. A map $\phi : G_1 \to G_2$ is an *anti-isomorphism* if it is a bijection and if both $\phi$ and $\phi^{-1}$ are anti-homomorphisms.

and is thus the subset of the set $M(n,\mathbb{R})$ of all real $n \times n$ matrices made up of those matrices that have an inverse.

The space $M(n,\mathbb{R})$ is in bijective correspondence with $\mathbb{R}^{n^2}$, and therefore it has a natural topological and differential structure. Note that the map $\det : M(n,\mathbb{R}) \to \mathbb{R}$, $A \mapsto \det A$, is continuous and hence $\det^{-1}\{0\}$ is a closed subset of $M(n,\mathbb{R})$ (since $\{0\}$ is a closed subset of $\mathbb{R}$). The subset $GL(n,\mathbb{R})$ is the complement of this set, and is hence open. As such, it acquires both a topological and a differentiable structure from $M(n,\mathbb{R})$, and it is with respect to this structure that it is a Lie group. Note that, as an open subset of the Euclidean space $\mathbb{R}^{n^2}$, the group $GL(n,\mathbb{R})$ has a non-compact topology.

Since $GL(n,\mathbb{R})$ is an open subset of $M(n,\mathbb{R}) \simeq \mathbb{R}^{n^2}$, the dimension of $GL(n,\mathbb{R})$ is the same as that of $M(n,\mathbb{R})$—namely $n^2$—and it decomposes into two disjoint components, according to whether $\det A$ is greater than, or less than, 0.

5. Note that, although it is not compact, the group $GL(n,\mathbb{R})$ is *locally compact*, *i.e.*, each point has a neighbourhood whose closure is compact. This property is important in many ways: for example, the existence of a measure on a Lie group that is quasi-invariant under left or right translations is only guaranteed if the group is locally compact, in which case one has the famous *Haar* measure. One of the major problems with infinite-dimensional Lie groups is that they are not locally compact, and hence the Haar measure does not exist. In particular, this poses severe difficulties for the quantisation of any system that possesses such a group: a good example is the problem of quantum gravity, and the role of the infinite-dimensional group of space-time, or spatial, diffeomorphisms.

6. The *complex, general linear group* in $n$-dimensions is defined as

$$GL(n,\mathbb{C}) := \{A \in M(n,\mathbb{C}) \mid \det A \neq 0\}. \qquad (4.1.9)$$

As an open subset of the $2n^2$ (real) dimensional manifold $M(n,\mathbb{C})$, it has the structure of a, non-compact, locally-compact, $2n^2$-dimensional Lie group.

Note that—unlike $GL(n,\mathbb{R})$—$GL(n,\mathbb{C})$ is a connected space since any matrix $A$ with $\det A > 0$ can be connected continuously to a matrix $B$ for which $\det B < 0$ by a path in $GL(n,\mathbb{C})$ on which the determinant is a complex number. Put another way, removing the single point 0 from the real line $\mathbb{R}$ disconnects it into two pieces whereas the same operation performed on $\mathbb{C}$ leaves a connected space.

7. The *connected component* of $GL(n,\mathbb{R})$ (*i.e.*, the component containing the unit element $\mathbb{1}$) is denoted

$$GL^+(n,\mathbb{R}) := \{A \in GL(n,\mathbb{R}) \mid \det A > 0\} \tag{4.1.10}$$

and is itself a non-compact, locally-compact, Lie group of dimension $n^2$.

To show directly that $GL^+(n,\mathbb{R})$ is a subgroup of $GL(n,\mathbb{R})$ note that:

(i) $\det(\mathbb{1}) = 1$ so that the unit element $\mathbb{1}$ belongs to $GL^+(n,\mathbb{R})$;

(ii) $\det AB = \det A \det B$ and hence $A$, $B \in GL^+(n,\mathbb{R})$ implies $AB \in GL^+(n,\mathbb{R})$;

(iii) $\det A^{-1} = (\det A)^{-1}$ and hence $A \in GL^+(n,\mathbb{R})$ implies $A^{-1} \in GL^+(n,\mathbb{R})$;

which are precisely the three conditions for a subset of a group $G$ to be a subgroup.

8. The *special linear groups* are defined as the following non-compact, locally compact, subgroups of the general linear groups:

$$SL(n,\mathbb{R}) := \{A \in GL(n,\mathbb{R}) \mid \det A = 1\} \tag{4.1.11}$$

and

$$SL(n,\mathbb{C}) := \{A \in GL(n,\mathbb{C}) \mid \det A = 1\}. \tag{4.1.12}$$

□

### 4.1.2 The orthogonal group

A very important role in theoretical physics is played by the orthogonal groups and the unitary groups.

**Definition 4.2**

The *(real) orthogonal group* is the subgroup of $GL(n, \mathbb{R})$ defined by

$$O(n, \mathbb{R}) := \{A \in GL(n, \mathbb{R}) \mid AA^t = \mathbb{1}\}. \tag{4.1.13}$$

**Comments**

1. To see that this subset of $GL(n, \mathbb{R})$ is indeed a subgroup note that:

(i) $\mathbb{1}\mathbb{1}^t = \mathbb{1}$, and hence the unit element belongs to $O(n, \mathbb{R})$;

(ii) if $AA^t = \mathbb{1}$ and $BB^t = \mathbb{1}$ then $(AB)^t(AB) = B^tA^tAB = \mathbb{1}$ and hence $O(n, \mathbb{R})$ is algebraically closed under multiplication;

(iii) if $AA^t = \mathbb{1}$ then $A^{-1} = A^t$ which means $A^{-1}(A^{-1})^t = A^tA = A^{-1}A = \mathbb{1}$ and hence $A \in O(n, \mathbb{R})$ implies that $A^{-1} \in O(n, \mathbb{R})$.

2. In terms of the matrix elements of the members of $O(n, \mathbb{R})$, the defining constraint $AA^t = \mathbb{1}$ is equivalent to the equations

$$\sum_{k=1}^{n} A_{ik}A_{jk} = \delta_{ij} \quad i, j = 1, 2, \ldots, n \tag{4.1.14}$$

which impose $(n^2 + n)/2$ polynomial constraints on the matrices regarded as elements of $M(n, \mathbb{R})$.

3. The definition of $O(n, \mathbb{R})$ in terms of the $(n^2 + n)/2$ polynomial constraints Eq. (4.1.14) plus the fact that $\dim GL(n, \mathbb{R}) = n^2$, suggests that $O(n, \mathbb{R})$ is a submanifold of $GL(n, \mathbb{R})$ of dimension

$n^2 - (n^2+n)/2 = n(n-1)/2$, and a more careful investigation shows that this is indeed so.[3]

However, even without embarking on such a detailed analysis, there are important pieces of information that can be extracted from the general form of Eq. (4.1.14). For example, consider the chain of maps

$$\begin{array}{ccccc} GL(n,\mathbb{R}) & \xrightarrow{\Delta} & GL(n,\mathbb{R})\times GL(n,\mathbb{R}) & \xrightarrow{\mathrm{id}\times T} & GL(n,\mathbb{R})\times GL(n,\mathbb{R}) \\ A & \mapsto & (A,A) & \mapsto & (A,A^t) \\ & & & \xrightarrow{\mu} & GL(n,\mathbb{R}) \\ & & & \mapsto & AA^t \end{array} \tag{4.1.15}$$

where $T: GL(n,\mathbb{R}) \to GL(n,\mathbb{R})$ is defined by $T(A) := A^t$, and $\mu$ is the multiplication function $\mu: GL(n,\mathbb{R}) \times GL(n,\mathbb{R}) \to GL(n,\mathbb{R})$, $\mu(A,B) := AB$. Now the composite function $\rho := \mu \circ (\mathrm{id} \times T) \circ \Delta$ is a continuous map from $GL(n,\mathbb{R})$ into itself with the property that

$$O(n,\mathbb{R}) = \rho^{-1}\{\mathbb{1}\}. \tag{4.1.16}$$

But the inverse image of a closed set (such as $\{\mathbb{1}\}$) under a continuous map is itself a closed set, and hence $O(n,\mathbb{R})$ is a *closed* submanifold of $GL(n,\mathbb{R})$. Furthermore, Eq. (4.1.14) implies that

$$\sum_{i=1}^{n}\sum_{j=1}^{n}(A_{ij})^2 = n \tag{4.1.17}$$

so that $O(n,\mathbb{R})$ is actually a closed subset of the sphere $S^{n^2-1} \subset M(n,\mathbb{R}) \cong \mathbb{R}^{n^2}$. In turn, this means that $O(n,\mathbb{R})$ is a closed and bounded subset of the vector space $M(n,\mathbb{R}) \cong \mathbb{R}^{n^2}$ and hence—by the Heine-Borel theorem (see Section 1.4.8)—it follows that the Lie group $O(n,\mathbb{R})$ is a *compact* manifold. This has major implications in many different directions; for example, the representation theory of a compact Lie group is far more tractable than is that of a noncompact group.

[3]This is not trivial since, in general, the solution surface of a collection of polynomial constraints has singular point. For example, see Postnikov (1986).

4. The condition $AA^t = \mathbb{1}$ implies that $(\det A)^2 = 1$, so that $\det A = \pm 1$. The closed (and hence[4] compact) subset of $O(n, \mathbb{R})$ of matrices $A$ satisfying $\det A = 1$ is a subgroup [Exercise!] known as the *special orthogonal group*, and is denoted $SO(n, \mathbb{R})$. It has dimension $n(n-1)/2$.

5. In an analogous way the *unitary* group $U(n)$ is defined by

$$U(n) := \{A \in GL(n, \mathbb{C}) \mid AA^\dagger = \mathbb{1}\} \tag{4.1.18}$$

and is a compact Lie subgroup of $GL(n, \mathbb{C})$ with real dimension $n^2$.

The *special unitary group* $SU(n)$ is defined as

$$SU(n) := \{A \in U(n) \mid \det A = 1\} \tag{4.1.19}$$

and is a compact Lie group of real dimension $n^2 - 1$.

6. The definition of the orthogonal group has an important generalisation. Namely, if the vector space $\mathbb{R}^n$ is equipped with the diagonal metric $(+1, \ldots, +1, -1, \ldots -1)$, with $p$ '+1' entries and $q$ '−1' entries—and such that $p + q = n$—then the subset of $GL(n, \mathbb{R})$ that leaves this metric fixed is a Lie group, and is denoted $O(p, q; \mathbb{R})$; the subgroup of matrices satisfying $\det A = 1$ is denoted $SO(p, q; \mathbb{R})$. The orthogonal group $O(n, \mathbb{R})$ introduced above corresponds to the special case of $O(p, q, \mathbb{R})$ with $p = n$, $q = 0$ (or $q = n$, $p = 0$). A similar remark applies to the special orthogonal groups.

Apart from $O(n, \mathbb{R})$ and $SO(n, \mathbb{R})$, all the groups $O(p, q; \mathbb{R})$ and $SO(p, q; \mathbb{R})$ are non-compact, locally-compact Lie groups. A case of particular importance in theoretical physics is $O(3, 1)$—the *Lorentz group* that plays a major role in the physics of special relativity.

# 4.2 The Lie Algebra of a Lie Group

## 4.2.1 Left-invariant vector fields

One of the most important features of a Lie Group $G$ is the existence of an associated Lie algebra that encodes many of the properties of

[4] In general, a closed subset of any compact space is itself compact.

the group. Rather remarkably, this even includes certain global topological properties of $G$ such as—for example—whether or not it is a compact space. The crucial property of a Lie group that enables this to occur is the existence of the left and right translations of $G$ onto itself that can be employed to map local, tangent-space related structure around the entire group manifold. The key definition is the following.

**Definition 4.3**

1. A vector field $X$ on a Lie group $G$ is *left-invariant* if it is $l_g$-related to itself for all $g \in G$ (see Eq. (3.1.25)), *i.e.*,

$$l_{g*}X = X \quad \text{for all } g \in G \tag{4.2.1}$$

or, equivalently,

$$l_{g*}(X_{g'}) = X_{gg'} \quad \text{for all } g, g' \in G. \tag{4.2.2}$$

2. Similarly, $X$ is *right-invariant* if it is $r_g$-related to itself for all $g \in G$, *i.e.*,

$$r_{g*}X = X \quad \text{for all } g \in G \tag{4.2.3}$$

or, equivalently,

$$r_{g*}(X_{g'}) = X_{g'g} \quad \text{for all } g, g' \in G. \tag{4.2.4}$$

**Comments**

1. The set of all left-invariant vector fields on a Lie group $G$ is denoted $L(G)$. It is clearly a real vector space.

2. The critical observation now is based on the general fact (Eq. (3.1.25)) that if vector fields $X_1$ and $X_2$ on a manifold $\mathcal{M}$ are $h$-related to vector fields $Y_1$ and $Y_2$ respectively on a manifold $\mathcal{N}$, where $h : \mathcal{M} \to \mathcal{N}$, then the commutator $[X_1, X_2]$ is $h$-related to the commutator $[Y_1, Y_2]$. In particular, this means that if $X_1$ and $X_2$ are left-invariant vector fields on $G$ then

$$\begin{aligned} l_{g*}[X_1, X_2] &= [l_{g*}X_1, l_{g*}X_2] \qquad \text{(from Eq. (3.1.25))} \\ &= [X_1, X_2]. \end{aligned} \tag{4.2.5}$$

Thus $X_1, X_2 \in L(G)$ implies that $[X_1, X_2] \in L(G)$, and so the set $L(G)$ of all left-invariant vector fields on $G$ is a sub Lie algebra of the infinite-dimensional Lie algebra of all vector fields on the manifold $G$. It is called the *Lie algebra* of $G$.

3. This result is very elegant, but it leaves open the question of whether there *are* any non-trivial left-invariant vector fields on any given Lie group $G$; and if such fields do exist, how many of them are there? Or to be more precise, what is the dimension of the real Lie algebra $L(G)$? The answer to these vital questions is given comprehensively by the theorem that follows. □

**Theorem 4.1** *There is an isomorphism of $L(G)$, regarded as a real vector space, with the tangent space $T_eG$ to $G$ at the unit element $e \in G$.*

**Proof**

An explicit isomorphism $i : T_eG \to L(G)$ will be constructed by exploiting the left-translation operation. Specifically, if $A \in T_eG$ let $i(A) := L^A$ where $L^A$ is the vector field on $G$ defined by

$$L^A_g := l_{g*}A \quad \text{for all } g \in G. \tag{4.2.6}$$

Note that, for all $g, g' \in G$,

$$l_{g'*}(L^A_g) = l_{g'*}(l_{g*}A) = l_{g'g*}A = L^A_{g'g} \tag{4.2.7}$$

where we have used $l_{g'} \circ l_g = l_{g'g}$, and hence $l_{g'*}l_{g*} = l_{g'g*}$. It follows from Eq. (4.2.7) that $L^A$ is indeed a left-invariant vector field.

It is clear that $A \to L^A$ is a linear map, and so it remains to prove that it is one-to-one and onto.

(i) If $L^A = L^B$ then, in particular, $L^A_e = L^B_e$. But $L^A_e = l_{e*}A = A$, and similarly $L^B_e = B$. Hence the map $i : T_eG \to L(G)$, $A \mapsto L^A$, is injective.

(ii) Now let $L$ be any left-invariant vector field on $G$, and define $A_L \in T_eG$ by

$$A_L := L_e \qquad [= l_{g^{-1}}{}_* L_g \text{ for all } g \in G]. \tag{4.2.8}$$

Clearly, $l_{g*}A_L = l_{g*}(L_e) = L_g$ since $L$ is left-invariant. Thus $i(A_L) = L$, and hence $i : T_eG \to L(G)$ is surjective. **QED**

**N.B.** It is clear that $L \mapsto A_L$ is the inverse of $A \mapsto L^A$, and hence

$$l_{g*}(A_L) = L_g \quad \text{and} \quad l_{g^{-1}}{}_*(L^A_g) = A. \tag{4.2.9}$$

**Corollary**

The dimension of the vector space $L(G)$ is equal to $\dim T_eG = \dim G$.

**Comments**

1. Since, by assumption, $G$ is a finite-dimensional manifold it follows that $L(G)$ is a finite-dimensional, nontrivial subalgebra of the Lie algebra of all vector fields on $G$.

2. A significant implication of the theorem is that it should be possible to regard the tangent space $T_eG$ as the Lie algebra of the Lie group $G$. However, the idea of a commutator is defined only for vector fields—not for vectors at a fixed point in $G$—and hence the Lie bracket on $T_eG$ has to be constructed from the commutator on the left-invariant vector fields $L(G)$ with the aid of the isomorphism $i : T_eG \to L(G)$ discussed in the theorem.

Specifically, if $A, B \in T_eG$ then the Lie bracket $[AB] \in T_eG$ is defined to be the unique element in $T_eG$ such that

$$L^{[AB]} = [L^A, L^B] \tag{4.2.10}$$

or, equivalently,

$$[AB] := [L^A, L^B]_e. \tag{4.2.11}$$

The power and elegance of this way of viewing the Lie algebra of a Lie group is reflected nicely in the following result.

**Theorem 4.2** *Let $f : G \to H$ be a smooth homomorphism between the Lie groups $G$ and $H$. Then the induced map $f_* : T_eG \to T_eH$ is a homomorphism between the Lie algebras of the groups.*

**Proof**

Since $f$ is a group homomorphism it follows that, for all $g, g' \in G$, $f(gg') = f(g)f(g')$, and hence for all $g \in G$

$$f \circ l_g = l_{f(g)} \circ f. \tag{4.2.12}$$

Let $A$ be in $T_eG$ with the associated left-invariant vector field $L^A \in L(G)$ (so that $A = L^A_e$) as defined in Eq. (4.2.6). Then

$$\begin{aligned} f_*(L^A_g) &= f_* l_{g*}(A) = (f \circ l_g)_*(A) = (l_{f(g)} \circ f)_*(A) \\ &= l_{f(g)*} f_*(A) = L^{f_*(A)}_{f(g)} \end{aligned} \tag{4.2.13}$$

which says that the vector fields $L^A$ and $L^{f_*(A)}$ are $f$-related. But then, from Eq. (3.1.30),

$$\begin{aligned} f_*[AA'] &= f_*([L^A, L^{A'}]_e) = [f_*(L^A), f_*(L^{A'})]_{f(e)} \\ &= [L^{f_*(A)}, L^{f_*(A')}]_e = [f_*(A) f_*(A')] \end{aligned} \tag{4.2.14}$$

and hence $f_* : T_eG \to T_eH$ is a Lie algebra homomorphism. **QED**

**Comments**

1. If $\{E_1, E_2, \ldots, E_n\}$, $n = \dim G$, is a basis set for $L(G) \cong T_eG$, then the commutator of any pair of these vector fields must be a linear combination of them. Thus

$$[E_\alpha, E_\beta] = \sum_{\gamma=1}^{n} C_{\alpha\beta}{}^{\gamma}\, E_\gamma \tag{4.2.15}$$

for some set of real numbers $C_{\alpha\beta}{}^{\gamma}$ that are known as the *structure constants* of the Lie group/algebra.

2. A similar construction to the above can be carried out with a set of right-invariant vector fields associated with the elements of $T_eG$. The analogue of Eq. (4.2.6) is the definition

$$R^A_g := r_{g*} A \tag{4.2.16}$$

with the 'inverse' $A_R := R_e$ $(= r_{g^{-1}*}(R_g)$ for any $g \in G)$.

It can be shown that

$$\text{(a)} \quad [R^A, R^{A'}] = R^{[A'A]} \equiv R^{-[AA']} \tag{4.2.17}$$

$$\text{(b)} \quad [R^A, L^{A'}] = 0 \tag{4.2.18}$$

$$\text{(c)} \quad R^A_g = (\mathrm{Ad}_{g^{-1}})_* \, L^A_g \tag{4.2.19}$$

where the *adjoint map* $\mathrm{Ad}_g : G \to G$ is defined for each $g \in G$ by

$$\mathrm{Ad}_g(g') := gg'g^{-1}. \tag{4.2.20}$$

## 4.2.2 The completeness of a left-invariant vector field

The next result is of central importance in the theory of Lie groups and their associated Lie algebras. Recall that an integral curve $t \to \sigma^X(t)$ of a vector field $X$ is said to be 'complete' (see Section 3.2) if it is defined (or can be extended to a curve that is defined) for all values of the parameter $t$. As remarked in Section 3.2, every vector field on a compact manifold is complete, but there are plenty of examples of incomplete vector fields on a non-compact space. However, the key result in Lie group theory is that *every* left-invariant vector field on a Lie group $G$ is complete, even if $G$ is non-compact.

To prove this, a preliminary result is needed concerning the 'infinitesimal' form of the product group law on $G$. Thus let $\mu : G \times G \to G$ be the group product, so that $\mu(g_1, g_2) := g_1 g_2$. We wish to consider the induced tangent space map $\mu_* : T_{(g_1,g_2)}G \times G \to T_{g_1 g_2}G$ which will be analysed with the aid of the discussion in Section 2.3.5. In particular, there is a map $\chi$ from $T_{(g_1,g_2)}G \times G$ onto the space $T_{g_1}G \oplus T_{g_2}G$ through which the induced map $\mu_*$ factorises as (see Eq. (2.3.74))

$$T_{(g_1,g_2)}G \times G \xrightarrow{\chi} T_{g_1}G \oplus T_{g_2}G \xrightarrow{\tilde{\mu}} T_{g_1 g_2}G \quad \text{where } \mu_* = \tilde{\mu} \circ \chi. \tag{4.2.21}$$

From Eq. (2.3.75),

$$\tilde{\mu}(\alpha, \beta) = (\mu \circ i_{g_2})_* \alpha + (\mu \circ j_{g_1})_* \beta \tag{4.2.22}$$

where $\alpha \in T_{g_1}G$, $\beta \in T_{g_2}G$, $i_{g_2}(g) = (g, g_2)$ and $j_{g_1}(g) = (g_1, g) \in G \times G$. Thus $\mu \circ i_{g_2}(g) = \mu(g, g_2) = gg_2$ and $\mu \circ j_{g_1}(g) = \mu(g_1, g) = g_1 g$, which can be rewritten as

$$\mu \circ i_{g_2} = r_{g_2} \quad \text{and} \quad \mu \circ j_{g_1} = l_{g_1}. \tag{4.2.23}$$

Hence Eq. (4.2.22) becomes

$$\tilde{\mu}(\alpha, \beta) = r_{g_2 *}(\alpha) + l_{g_1 *}(\beta) \tag{4.2.24}$$

for all $(\alpha, \beta) \in T_{g_1}G \oplus T_{g_2}G$. This is the result that will be needed in the proof of the following main theorem.

**Theorem 4.3** *If $X$ is a left-invariant vector field on a Lie group $G$, then $X$ is complete.*

### Proof

1. Let $t \to \sigma^X(t)$ be an integral curve of $X$ through $e \in G$ that is defined for all $|t| < \epsilon$ for some $\epsilon > 0$. Thus

$$\sigma^X_* \left(\frac{d}{dt}\right)_s = X_{\sigma^X(s)} \quad \text{for all } |s| < \epsilon \tag{4.2.25}$$

and hence

$$(l_g \circ \sigma^X)_* \left(\frac{d}{dt}\right)_s = l_{g*}\sigma^X_* \left(\frac{d}{dt}\right)_s = l_{g*}(X_{\sigma^X(s)}) = X_{g\sigma^X(s)} \tag{4.2.26}$$

which shows that $t \to l_g \circ \sigma^X(t)$ is an integral curve of $X$ through the point $g \in G$.

It follows that if $\phi^X_t(g) := g\sigma^X(t)$, then $t \to \phi^X_t$ is a local flow (see Section 3.2) for the vector field $X$. This implies that $\phi^X_{t_1}(\phi^X_{t_2}(g)) = \phi^X_{t_1+t_2}(g)$ and so

$$\sigma^X(t_1 + t_2) = \sigma^X(t_1)\sigma^X(t_2) \tag{4.2.27}$$

for all $t_1, t_2$ such that $|t_1| < \epsilon$, $|t_2| < \epsilon$ and $|t_l + t_2| < \epsilon$. This equation can be read as saying that the map $t \to \sigma^X(t)$ is a local homomorphism from the additive group $\mathbb{R}$ of the real line into the Lie group $G$.

2. This suggests a means whereby the integral curves on $G$ can be extended consistently using the group structure. Namely, if $|t| < 2\epsilon$, define

$$\sigma^X(t) := \left(\sigma^X\left(\frac{t}{2}\right)\right)^2 \tag{4.2.28}$$

which is consistent with the original definition if $|t| < \epsilon$ since then, using Eq. (4.2.27),

$$\left(\sigma^X\left(\frac{t}{2}\right)\right)^2 = \sigma^X(t). \tag{4.2.29}$$

Clearly this procedure can be iterated to $|t| < n\epsilon$ for any positive integer $n$, and hence to the whole real line. Thus the crucial task is to show that the curve defined in (4.2.28) is, in fact, an integral curve of $X$.

3. Temporarily denoting the curve defined by the right hand side of Eq. (4.2.28) by $t \mapsto \tilde{\sigma}^X(t)$, it is necessary to show that for $|s| < 2\epsilon$, $\tilde{\sigma}^X_*\left(\frac{d}{dt}\right)_s = X_{\sigma^X(s)}$. To do this, factorise the expression (4.2.28) as $\tilde{\sigma}^X = \mu \circ \Delta \circ \sigma^X \circ H$ where $H : \mathbb{R} \to \mathbb{R}$ is defined by $H(t) := \frac{t}{2}$, and $\Delta : G \to G \times G$ is the usual diagonal map $\Delta(g) = (g, g)$. Thus

$$\begin{aligned}
\tilde{\sigma}^X_*\left(\frac{d}{dt}\right)_s = \frac{1}{2}\mu_* \Delta_* \sigma^X{}_*\left(\frac{d}{dt}\right)_{\frac{s}{2}} &= \frac{1}{2}\mu_* \Delta_*\left(X_{\sigma^X(s/2)}\right) \qquad (4.2.30)\\
&= \frac{1}{2}\tilde{\mu} \circ \chi \circ \Delta_*\left(X_{\sigma^X(s/2)}\right)
\end{aligned}$$

where the factorisation from Eq. (4.2.21) has been used. Then, from Eq. (2.3.79),

$$\begin{aligned}
\tilde{\sigma}^X_*\left(\frac{d}{dt}\right)_s &= \frac{1}{2}\tilde{\mu}(X_{\sigma^X(s/2)}, X_{\sigma^X(s/2)})\\
&= \frac{1}{2}\left\{r_{\sigma^X(s/2)_*}(X_{\sigma^X(s/2)}) + l_{\sigma^X(s/2)_*}(X_{\sigma^X(s/2)})\right\} \qquad (4.2.31)
\end{aligned}$$

where Eq. (4.2.24) has been used. However, the left and right translations of a group commute (*i.e.*, $r_g \circ l_g = l_g \circ r_g$ for all $g \in G$) and hence the first term on the right hand side of Eq. (4.2.31) can be written as (using also the fact that the vector field $X$ is left-invariant):

$$(r_{\sigma^X(s/2)} \circ l_{\sigma^X(s/2)})_* X_e = (l_{\sigma^X(s/2)} \circ r_{\sigma^X(s/2)})_* X_e. \tag{4.2.32}$$

But, from Eq. (4.2.27), $r_{\sigma^X(t)} \circ \sigma^X = l_{\sigma^X(t)} \circ \sigma^X$, and hence

$$r_{\sigma^X(t)*}X_e = (r_{\sigma^X(t)} \circ \sigma^X)_* \left(\frac{d}{dt}\right)_0 = \left(l_{\sigma^X(t)} \circ \sigma^X\right)_* \left(\frac{d}{dt}\right)_0$$
$$= l_{\sigma^X(t)*}X_e = X_{\sigma^X(t)}. \quad (4.2.33)$$

Thus the right hand side of Eq. (4.2.32) is $l_{\sigma^X(s/2)*}X_{\sigma^X(s/2)} = X_{(\sigma^X(s/2))^2}$, and this is the first term on the right hand side of Eq. (4.2.31). But the second term on the right hand side of this equation is clearly the same, and hence Eq. (4.2.31) reads

$$\tilde{\sigma}^X_* \left(\frac{d}{dt}\right)_s = X_{(\sigma^X(s/2))^2} = X_{\sigma^X(s)} \quad (4.2.34)$$

which is precisely the statement that $\tilde{\sigma}^X$ defined by Eq. (4.2.28) is an extension of the original integral curve $\sigma^X$ from the interval $|t| < \epsilon$ to $|t| < 2\epsilon$. Iterating the procedure then shows that the integral curve of $X$ can be extended to all values of $t \in \mathbb{R}$, and hence the left-invariant vector field $X$ is complete. **QED**

### 4.2.3 The exponential map

The theorem just proved leads naturally to the important idea of the 'exponential' map.

**Definition 4.4**

1. The unique integral curve $t \to \sigma^{L^A}(t)$, $A = \sigma^{L^A}_*\left(\frac{d}{dt}\right)_0$, of the left-invariant vector field $L^A$ (with $\sigma^{L^A}(0) = e$) that is defined for all $t \in \mathbb{R}$ by virtue of the above theorem is written as

$$t \mapsto \exp tA \quad (4.2.35)$$

where $A \in T_eG$.

2. The *exponential map* $\exp : T_eG \to G$ is defined by

$$\exp A := \exp tA\,|_{t=1}. \quad (4.2.36)$$

3. A *one-parameter subgroup* of a Lie group $G$ is a smooth homomorphism $\chi$ from the additive group of the real line $\mathbb{R}$ into $G$; *i.e.*, for all $t_1, t_2 \in \mathbb{R}$,

$$\chi(t_1 + t_2) = \chi(t_1)\chi(t_2). \tag{4.2.37}$$

**Comments**

1. It can be shown[5] that the exponential map is a local *diffeomorphism* from the tangent space $T_eG$ at the unit element $e \in G$ into $G$. Thus there is some open neighbourhood $V$ of $0 \in T_eG$ such that (i) the exponential map restricted to $V$ is bijective; and (ii) both it and its inverse (defined on the image of $V$ in $G$ under exp) are smooth.

2. If $G$ is a compact Lie group then the exponential map is a surjective function from $T_eG$ onto $G$. Of necessity it is then *not* injective as else it would establish a diffeomorphism between the non-compact vector space $T_eG$ and the compact group $G$. The statement in Comment 1 still applies however and hence the map *is* one-to-one in some neighbourhood of $e \in G$ (it is assumed here that $G$ is connected). A good example is the case $G = U(1)$, whose Lie algebra is simply $\mathbb{R}$. It can be shown that $\exp r = e^{ir}$, so that values of $r \in \mathbb{R}$ that differ by an integral multiple of $2\pi$ are mapped to the same point in $U(1)$.

If $G$ is not compact, then the exponential map may not be surjective (there are some famous simple examples[6] of this phenomenon). On the other hand, it *is* now possible for the map to be globally injective.

3. It follows from Eq. (4.2.27) (and its equivalent for the fully extended integral curve) that $t \mapsto \exp tA$ is a one-parameter subgroup of $G$. The following proposition shows that the converse is also true: namely, *every* one-parameter subgroup of $G$ is of the form $t \mapsto \exp tA$ for some $A \in T_eG \cong L(G)$. Thus there is a one-to-one association between one-parameter subgroups of the Lie group $G$ and its Lie

[5]A classic text for this type of result is Chevalley (1946); see also Helgason (1962).

[6]For example, see Helgason (1962).

algebra. The neighbourhood of $e \in G$ onto which $\exp : T_eG \to G$ maps diffeomorphically is 'filled' with the images of these subgroup maps.

**Theorem 4.4** *If* $\chi : \mathbb{R} \to G$ *is a one-parameter subgroup of* $G$ *then, for all* $t \in \mathbb{R}$, $\chi(t) = \exp tA$ *where* $A := \chi_* \left(\frac{d}{dt}\right)_0$.

**Proof**

The relation (4.2.37) implies that $\chi \circ l_s = l_{\chi(s)} \circ \chi$ for all $s \in \mathbb{R}$. Hence

$$\chi_* \left(\frac{d}{dt}\right)_s = \chi_* l_{s*} \left(\frac{d}{dt}\right)_0 = l_{\chi(s)_*} \chi_* \left(\frac{d}{dt}\right)_0 = l_{\chi(s)_*}(A) = L^A_{\chi(s)} \tag{4.2.38}$$

and hence $t \mapsto \chi(t)$ is an integral curve for $L^A \in L(G)$. The result follows from the uniqueness of such curves. **QED**

**Corollary**

Let $f : G \to H$ be a smooth homomorphism between two Lie groups $G$ and $H$ whose exponential maps are denoted $\exp_G : T_eG \to G$ and $\exp_H : T_eH \to H$ respectively. Then we have the commutative diagram

$$\begin{array}{ccc} T_eG & \xrightarrow{f_*} & T_eH \\ \downarrow{\scriptstyle \exp_G} & & \downarrow{\scriptstyle \exp_H} \\ G & \xrightarrow{f} & H \end{array} \tag{4.2.39}$$

*i.e.*, for all $A \in T_eG$, $\exp_H(f_* A) = f(\exp_G A)$.

**Proof**

Define $\chi : \mathbb{R} \to H$ by $\chi(t) := f(\exp_G tA)$. Then

$$\begin{aligned} \chi(t_1 + t_2) &= f(\exp_G(t_1 + t_2)A) = f(\exp_G t_1 A \exp_G t_2 A) \\ &= f(\exp_G t_1 A)\, f(\exp_G t_2 A) = \chi(t_1)\chi(t_2) \end{aligned} \tag{4.2.40}$$

so that $\chi$ is a one-parameter subgroup of $H$. It follows from the theorem that

$$\chi(t) = \exp_H tB \quad \text{where } B := \chi_* \left(\frac{d}{dt}\right)_o \in T_eH. \tag{4.2.41}$$

Let $k \in C^\infty(H)$. Then, from Eq. (2.3.40),

$$B(k) = \left(\frac{d}{dt}\right)_0 (k \circ \chi) = \left.\frac{d}{dt}k(\chi(t))\right|_{t=0} = \left.\frac{d}{dt}k \circ f(\exp_G tA)\right|_{t=0} = L_e^A(k \circ f) \tag{4.2.42}$$

where the last step follows because $t \mapsto \exp_G tA$ is an integral curve for the left-invariant vector field $L^A$. But $L_e^A = A$ and hence $B(k) = A(k \circ f) = (f_* A)(k)$, *i.e.*, $B = f_*(A)$. Thus Eq. (4.2.41) reads

$$f(\exp_G tA) = \chi(t) = \exp_H t f_*(A) \tag{4.2.43}$$

which, for $t = 1$, proves the result in Eq. (4.2.39). **QED**

### Corollary

If $\mathrm{Ad}_g(g') := g g' g^{-1}$ for each $g \in G$ then

$$\exp \mathrm{Ad}_{g*} B = g \exp B \, g^{-1} \quad \text{for all } B \in T_e G. \tag{4.2.44}$$

### Proof

We note that $\mathrm{Ad}_g(e) = e$, and hence $\mathrm{Ad}_{g_*}$ maps $T_eG$ to $T_eG$; hence Eq. (4.2.44) is well defined. However, for each $g \in G$, the mapping $\mathrm{Ad}_g : G \to G$ is a homomorphism of the Lie group onto itself. Therefore, from the corollary above,

$$\exp \mathrm{Ad}_{g*} B = \mathrm{Ad}_g(\exp B) = g \exp B \, g^{-1}. \tag{4.2.45}$$

**QED**

The map $g \mapsto \mathrm{Ad}_{g*}$ gives a representation of $G$ on $T_eG \cong L(G)$ known as the *adjoint* representation. Note that its kernel is the centre of $G$.

For example, for the Lie group $SU(3)$, the adjoint representation is eight dimensional, and the kernel of the action is $\mathbf{Z}_3$. This is the representation that is used in the famous 'eight-fold way' classification of the elementary particles.

### 4.2.4 The Lie algebra of $GL(n, \mathbb{R})$

An important special example of left-invariant vector fields is when the Lie group concerned is the connected component $GL^+(n, \mathbb{R})$ of the general linear group $GL(n, \mathbb{R})$. This group appears as an open subset of the linear space $M(n, \mathbb{R})$ of all real $n \times n$ matrices: hence the tangent space at any point in $G$ (and especially at $e \in G$) can be identified in a natural way with the vector space $M(n, \mathbb{R})$, which can therefore in turn be associated with the Lie algebra of $GL^+(n, \mathbb{R})$.

A natural system of coordinates on $GL^+(n, \mathbb{R})$—valid in some neighbourhood of the unit matrix $\mathbb{1}$—are the 'matrix elements' defined by

$$x^{ij}(g) := g^{ij} \tag{4.2.46}$$

where $g \in GL^+(n, \mathbb{R})$ and $i, j = 1, 2, \ldots, n$. Let $A \in T_eG \cong M(n, \mathbb{R})$. Then the coordinate representation of the left-invariant vector field associated with $A$ is

$$L_g^A = \sum_{i,j=1}^{n} \left(L^A x^{ij}\right)_g \left(\frac{\partial}{\partial x^{ij}}\right)_g \tag{4.2.47}$$

and

$$\left(L^A x^{ij}\right)_g = \frac{d}{dt}\left(x^{ij}(g\,\exp tA)\right)_{t=0}. \tag{4.2.48}$$

However, $A \in M(n, \mathbb{R})$ is a matrix and hence we can consider the curve $t \mapsto e^{tA}$ in $GL^+(n, \mathbb{R})$ where $e^{tA}$ refers to the usual exponential of a matrix. But the tangent vector to this curve at $t = 0$ is obviously the matrix $A$ and, furthermore, it is clear that it defines a one-parameter subgroup of $GL^+(n, \mathbb{R})$. Hence

$$e^{tA} = \exp tA \tag{4.2.49}$$

for all $t \in \mathbb{R}$ and $A \in T_eG \cong M(n, \mathbb{R})$. In particular, Eq. (4.2.47) can be rewritten using

$$\begin{aligned} L_g^A x^{ij} = \frac{d}{dt} x^{ij}(g\, e^{tA})\bigg|_{t=0} &= \sum_{k=1}^{n} g^{ik} \frac{d}{dt}\left(e^{tA}\right)^{kj}\bigg|_{t=0} \\ &= \sum_{k=1}^{n} g^{ik} A^{kj} = (gA)^{ij} \end{aligned} \tag{4.2.50}$$

so that the left-invariant vector field $L^A$ has the local-coordinate representation

$$L^A_g = \sum_{i,j=1}^{n} (gA)^{ij} \left(\frac{\partial}{\partial x^{ij}}\right)_g . \tag{4.2.51}$$

**Comments**

1. Similarly, the right-invariant vector fields on $GL^+(n, \mathbb{R})$ can be written in coordinate form as

$$R^A_g = \sum_{i,j=1}^{n} (Ag)^{ij} \left(\frac{\partial}{\partial x^{ij}}\right)_g . \tag{4.2.52}$$

2. The representation (4.2.51) gives

$$[L^{A'}, L^A] = L^{[A',A]} \tag{4.2.53}$$

where $[A', A]$ is the usual matrix commutator: hence the Lie algebra structure induced on $T_e GL^+(n, \mathbb{R}) \cong M(n, \mathbb{R})$ is just the commutator of the matrices.

3. A natural basis for $M(n, \mathbb{R})$ is the set of matrices $E_{ij}$ defined as

$$(E_{ij})_{kl} := \delta_{ik}\delta_{jl} \tag{4.2.54}$$

and the associated left-invariant vector fields are

$$L^{ij}_g = \sum_{k=1}^{n} g^{ki} \left(\frac{\partial}{\partial x^{kj}}\right)_g . \tag{4.2.55}$$

## 4.3 Left-Invariant Forms

### 4.3.1 The basic definitions

The concept of left-invariance for vector fields on a Lie group $G$ has a natural analogue with respect to the dual structure of differential forms. Two of the most important applications for left-invariant forms are (i) the development of a group-invariant DeRham cohomology for Lie groups; and (ii) the theory of connections in fibre bundles. The latter topic has been much discussed by theoretical physicists in recent years in relation to the general mathematical framework for Yang-Mills theories, particularly in investigations of anomalies in the quantisation of such systems. The discussion that follows of invariant differential forms is not very comprehensive and is partly concerned with developing the minimal set of tools necessary to set up the theory of connections in bundles.

**Definition 4.5**

An $n$-form $\omega \in A^n(G)$ on a Lie group is *left-invariant* if, for all $g \in G$,

$$l_g^*\omega = \omega \tag{4.3.1}$$

or, equivalently,

$$l_g^*(\omega_{g'}) = \omega_{g^{-1}g'} \tag{4.3.2}$$

for all $g' \in G$.

A *right-invariant* differential form is defined in an analogous way.

**Comments**

1. From Eq. (3.4.18) follows $l_g^*(d\omega) = d(l_g^*\omega)$, and hence if $\omega$ is left-invariant so is its exterior derivative $d\omega$. This is one of the starting points for the development of a $G$-invariant version of the DeRham differential form cohomology.

2. It was shown in Theorem 4.1 that there is an isomorphism $i : T_eG \to L(G)$ in which $i(A) = L^A$. Similarly, there is an isomorphism between $T_e^*G$ and the set $L^*(G)$ of left-invariant one-forms on $G$ that associates to each $d \in T_e^*G$ the left-invariant one-form $\lambda^d$ on $G$ defined by

$$\lambda_g^d := l_{g^{-1}}{}^*(d) \in T_g^*G \tag{4.3.3}$$

for all $g \in G$.

In this context, $T_e^*G \cong L^*(G)$ is called the *dual Lie algebra* of $G$. However, this nomenclature is potentially confusing in so far as $L^*(G)$

is *not* itself a Lie algebra—it is only the linear dual of the underlying vector space of the Lie algebra $L(G)$.

3. The left-invariant one-forms defined in Eq. (4.3.3) stand in a dual relationship to the left-invariant vector fields $L^A$ according to the equation

$$\langle \lambda^d, L^A \rangle_g = \langle d, A \rangle \text{ for all } g \in G \tag{4.3.4}$$

where $\langle d, A \rangle$ refers to the pairing between $A \in T_eG$ and $d$ as an element of the algebraic dual vector space $T_e^*G$. □

### 4.3.2 The Cartan-Maurer form

In Eq. (4.2.15) we introduced the idea of structure constants with respect to a particular basis $\{E_1, E_2, \ldots, E_n\}$, $n = \dim G$, of left-invariant vector fields:

$$[E_\alpha, E_\beta] = \sum_{\gamma=1}^{n} C_{\alpha\beta}{}^{\gamma} E_\gamma. \tag{4.3.5}$$

A dual basis $\{\omega^1, \omega^2, \ldots, \omega^n\}$ for $L^*(G)$ can be defined by

$$\langle \omega^\alpha, E_\beta \rangle := \delta^\alpha_\beta \tag{4.3.6}$$

but—in seeking to derive an analogue of Eq. (4.3.5)—it should be noted that, whereas the commutator of two vector fields is itself a vector field, the natural combination of a pair of one-forms (with a wedge-product) leads not to a one-form but to a two-form. However, a two-form can also be reached from a one-form by exterior differentiation, which suggests that both this and the operation of taking a wedge-product may be involved in the analogue of Eq. (4.3.5) for a basis of left-invariant one-forms.

This intuition can be substantiated by examining the two-forms $d\omega^\alpha$ via their effect on a pair of the basis elements for $L(G)$ using Eq. (3.4.15):

$$d\omega^\alpha(E_\beta, E_\gamma) = E_\beta(\langle \omega^\alpha, E_\gamma \rangle) - E_\gamma(\langle \omega^\alpha, E_\beta \rangle) - \langle \omega^\alpha, [E_\beta, E_\gamma] \rangle. \tag{4.3.7}$$

But $\langle \omega^\alpha, E_\gamma \rangle = \delta^\alpha_\gamma$, which is a constant and is hence annihilated by the vector field $E_\beta$. Thus using Eq. (4.3.5) in the last term of Eq. (4.3.7) gives

$$d\omega^\alpha(E_\beta, E_\gamma) = -C_{\beta\gamma}{}^\alpha. \tag{4.3.8}$$

However, the basic definition of the wedge product in Eq. (3.4.10) gives

$$\begin{aligned} \omega^\delta \wedge \omega^\epsilon(E_\beta, E_\gamma) &= \omega^\delta \otimes \omega^\epsilon(E_\beta, E_\gamma) - \omega^\delta \otimes \omega^\epsilon(E_\gamma, E_\beta) \\ &= \delta^\delta_\beta \delta^\epsilon_\gamma - \delta^\delta_\gamma \delta^\epsilon_\beta \end{aligned} \tag{4.3.9}$$

which can be combined with Eq. (4.3.8) to give the famous *Cartan-Maurer* equation

$$d\omega^\alpha + \frac{1}{2} \sum_{\beta,\gamma=1}^{n} C_{\beta\gamma}{}^\alpha \omega^\beta \wedge \omega^\gamma = 0 \tag{4.3.10}$$

for the exterior derivative of a left-invariant one-form.

**Definition 4.6**

The *Cartan-Maurer* form $\Xi$ is the $L(G)$-valued one-form on $G$ that associates with any $v \in T_gG$ the left-invariant vector field on $G$ whose value at $g \in G$ is precisely the given tangent vector $v$.

Specifically, if $\langle \Xi, v \rangle$ denotes this left-invariant vector field then

$$\langle \Xi, v \rangle(g') := l_{g'*}(l_{g^{-1}*}v) \tag{4.3.11}$$

for all $v \in T_gG$.

**Comments**

1. The Cartan-Maurer form is left-invariant.

2. On the left-invariant vector fields $L^A$, the expression Eq. (4.3.11) becomes

$$\langle \Xi, L^A_g \rangle(g') = L^A_{g'}. \tag{4.3.12}$$

3. Since $L(G) \cong T_eG$ it is also possible (and perhaps potentially less confusing) to regard the Cartan-Maurer form as taking its values

in $T_eG$, rather than in $L(G)$. With this interpretation Eq. (4.3.12) becomes

$$\langle \Xi, L^A_g \rangle = A \tag{4.3.13}$$

which, in fact, serves to define $\Xi$ precisely.

In particular, when $G = GL(n, \mathbb{R})$ with $T_eG \cong M(n, \mathbb{R})$, $\Xi$ can be viewed as an $M(n, \mathbb{R})$-valued (*i.e.*, a matrix-valued) differential form on the group. From Eq. (4.3.13) and the explicit expression Eq. (4.2.51) for $L^A$ in the case of $GL(n, \mathbb{R})$, it follows that

$$\Xi^{ij}_g = \sum_{k=1}^{n} (g^{-1})^{ik} (dx^{kj})_g. \tag{4.3.14}$$

4. Let $\Omega : \mathcal{M} \to G$ where $\mathcal{M}$ is some differentiable $m$-dimensional manifold: for example, if $\mathcal{M}$ is space-time or physical space, $\Omega$ could be thought of as a 'gauge-function' associated with a Yang-Mills theory using the Lie group $G$. Then $\Omega^*\Xi$ is a $L(G)$-valued one-form on $\mathcal{M}$. When $G$ is a group of matrices, Eq. (4.3.14) can be used to write the components of $\Omega^*\Xi$ with respect to some local coordinate system on $G$ as

$$\begin{aligned}
\langle (\Omega^*\Xi)^{ij}_p, \left(\frac{\partial}{\partial x^\mu}\right)_p \rangle &= \langle \Xi^{ij}, \Omega_* \left(\frac{\partial}{\partial x^\mu}\right) \rangle_{\Omega(p)} \\
&= \langle \sum_{k=1}^{n} \left(\Omega^{-1}(p)\right)^{ik} (dx^{kj})_{\Omega(p)}, \Omega_* \left(\frac{\partial}{\partial x^\mu}\right) \rangle_{\Omega(p)} \\
&= \sum_{k=1}^{n} \left(\Omega^{-1}(p)\right)^{ik} \Omega_* \left(\frac{\partial}{\partial x^\mu}\right)_p (x^{kj}) \\
&= \sum_{k=1}^{n} \left(\Omega^{-1}(p)\right)^{ik} \frac{\partial}{\partial x^\mu} x^{kj}(\Omega(p))
\end{aligned} \tag{4.3.15}$$

for all $p \in \mathcal{M}$ in the coordinate chart. Hence

$$(\Omega^*\Xi)^{ij}_p = \sum_{\mu=1}^{m} \sum_{k=1}^{n} \left(\Omega^{-1}(p)\right)^{ik} \frac{\partial}{\partial x^\mu} \Omega^{kj}(p)\, (dx^\mu)_p, \tag{4.3.16}$$

which is often written rather symbolically as

$$\Omega^*\Xi = \Omega^{-1} d\Omega. \tag{4.3.17}$$

□

**Exercises**

Show that under the action $g \mapsto r_g$ of right translation of $G$ onto itself, the Cartan-Maurer form transforms as

$$r_{g*}(\Xi) = \mathrm{Ad}_{g^{-1}*}(\Xi) \tag{4.3.18}$$

in the sense that, for any $v \in T_{g'}G$, $\langle r_{g*}(\Xi), v\rangle$ is the element of $T_eG$ given by $\mathrm{Ad}_{g^{-1}*}(\langle \Xi, v\rangle)$.

# 4.4 Transformation Groups

## 4.4.1 The basic definitions

In almost all applications of group theory in theoretical physics the groups arise as groups of transformations of some space of physical significance; for example, this could be space-time, or space, or some set of references frames, or something somewhat more exotic. Thus the idea of transformation groups is of fundamental importance in theoretical physics. This motivates the following sequence of definitions.

**Definition 4.7**

A group $G$ *acts* on a set $\mathcal{M}$ *on the left*, or *has a left action* on $\mathcal{M}$, if there is a homomorphism $g \mapsto \gamma_g$ from $G$ into the group $\mathrm{Perm}(\mathcal{M})$ of bijections of $\mathcal{M}$. Thus

$$\text{(i)} \quad \gamma_e(p) = p \text{ for all } p \in \mathcal{M}; \tag{4.4.1}$$

$$\text{(ii)} \quad \gamma_{g_2} \circ \gamma_{g_1} = \gamma_{g_2 g_1} \text{ for all } g_1, g_2 \in G. \tag{4.4.2}$$

**Comments**

1. A linear representation of a group $G$ is a special example of a $G$-action in which the set on which the group acts is a vector space, and with the action respecting the linear structure.

2. The image point $\gamma_g(p)$ is usually written as simply $gp$, in which case equations (4.4.1) and (4.4.2) become

$$\text{(i)} \quad ep = p \ \text{ for all } p \in \mathcal{M}; \tag{4.4.3}$$

$$\text{(ii)} \quad g_2(g_1 p) = (g_2 g_1)p \ \text{ for all } g_1, g_2 \in G. \tag{4.4.4}$$

3. There is an analogous definition of a *right action* of a group $G$ on a set $\mathcal{M}$. This is an *anti*-homomorphism $g \mapsto \delta_g$ from $G$ into $\text{Perm}(\mathcal{M})$. Thus

$$\text{(i)} \quad \delta_e(p) = p \ \text{ for all } p \in \mathcal{M}; \tag{4.4.5}$$

$$\text{(ii)} \quad \delta_{g_2} \circ \delta_{g_1} = \delta_{g_1 g_2} \ \text{ for all } g_1, g_2 \in G. \tag{4.4.6}$$

Writing the image point $\delta_g(p)$ in this case as $pg$, equations (4.4.5) and (4.4.6) become

$$\text{(i)} \quad pe = p \ \text{ for all } p \in \mathcal{M}; \tag{4.4.7}$$

$$\text{(ii)} \quad (pg_1)g_2 = p(g_1 g_2) \ \text{ for all } g_1, g_2 \in G. \tag{4.4.8}$$

4. If $\mathcal{M}$ is a topological space (resp. differentiable manifold) then usually it is appropriate to require that the bijections involved are restricted to the subgroup of $\text{Perm}(\mathcal{M})$ of homeomorphisms (resp. diffeomorphisms) of $\mathcal{M}$.

However, if $G$ is a topological (resp. Lie) group then it is natural to try to put a topological (resp. differential) structure on the group of homeomorphisms (resp. diffeomorphisms) of $\mathcal{M}$ and then to require that the maps $g \mapsto \gamma_g$ or $g \mapsto \delta_g$ be continuous (resp. differentiable) with respect to this new structure.

Such a step is not impossible, but it is complicated since these groups are infinite-dimensional and hence necessarily involve the non-trivial subtleties of infinite-dimensional topology and differential geometry. This seems an unnecessary complication, especially since most of the Lie groups $G$ with which theoretical physicists are concerned are themselves[7] finite-dimensional. The resolution of this problem is contained in the definition that follows, which uses the same

[7]Important exceptions are the group of diffeomorphisms of space, or space-time, and the group of gauge transformations in Yang-Mills theory.

type of trick discussed earlier in the context of Eq. (3.2.11) and the theory of flows of vector fields. □

**Definition 4.8**

A *left action* of a Lie group $G$ on a differentiable manifold $\mathcal{M}$ is a homomorphism $g \mapsto \gamma_g$ from $G$ into the group of diffeomorphisms Diff($\mathcal{M}$) of $\mathcal{M}$ with the property that the map $\Gamma : G \times \mathcal{M} \to \mathcal{M}$ defined by

$$\begin{aligned} \Gamma : G \times \mathcal{M} &\to \mathcal{M} \\ (g, p) &\mapsto \gamma_g(p) \equiv gp \end{aligned} \tag{4.4.9}$$

is a smooth map from the differentiable manifold $G \times \mathcal{M}$ into the manifold $\mathcal{M}$.

**Comments**

1. As remarked earlier, a special case of a left action is a linear representation of a Lie group on a finite-dimensional vector space. This is required to be smooth in the sense above.

2. The definition of a left action is often given directly in terms of this map $\Gamma : G \times \mathcal{M} \to \mathcal{M}$, in which case the basic homomorphism conditions in equations (4.4.1) and (4.4.2) become

$$\text{(i)} \quad \Gamma(e, p) = p \quad \text{for all } p \in \mathcal{M} \tag{4.4.10}$$

$$\text{(ii)} \quad \Gamma(g_2, \Gamma(g_1, p)) = \Gamma(g_2 g_1, p) \quad \text{for all } g_1, g_2 \in G. \tag{4.4.11}$$

3. Whenever one encounters a map between two manifolds, it is always interesting to look at the associated push-forward operation on the tangent spaces. In the present case, this suggests we should study the map

$$\Gamma_* : T_{(g,p)} G \times \mathcal{M} \to T_{gp} \mathcal{M}, \tag{4.4.12}$$

and we shall essentially do so in Section 4.5.

4. There is, of course, an equivalent definition for a right action of a Lie group on a differentiable manifold. Note that, in practice,

there is no need to consider left-actions and right-actions separately since they are in one-to-one correspondence. Specifically, given a left action $g \mapsto \gamma_g$ a right action can be defined by

$$\delta_g := \gamma_{g^{-1}} \tag{4.4.13}$$

for all $g \in G$, and vice versa.

5. There is a well-defined concept of a structure-preserving map between a pair of group actions of a Lie group $G$ on manifolds $\mathcal{M}$ and $\mathcal{M}'$. Namely, a map $f : \mathcal{M} \to \mathcal{M}'$ is said to be *equivariant* with respect to the group actions $g \mapsto \gamma_g$ and $g \mapsto \gamma'_g$ on $\mathcal{M}$ and $\mathcal{M}'$ respectively if the following diagram commutes

$$\begin{array}{ccc} \mathcal{M} & \xrightarrow{f} & \mathcal{M}' \\ \downarrow{\scriptstyle \gamma_g} & & \downarrow{\scriptstyle \gamma'_g} \\ \mathcal{M} & \xrightarrow{f} & \mathcal{M}' \end{array} \tag{4.4.14}$$

so that

$$\gamma'_g \circ f = f \circ \gamma_g \tag{4.4.15}$$

for all $g \in G$. Note that Eq. (4.4.15) is equivalent to the statement

$$gf(p) = f(gp) \tag{4.4.16}$$

for all $g \in G$ and $p \in \mathcal{M}$.[8]

6. A more sophisticated version of this idea allows for two Lie groups $G$ and $G'$ that act on manifolds $\mathcal{M}$ and $\mathcal{M}'$ respectively, and with a homomorphism $\rho : G \to G'$. In this case, a map $f : \mathcal{M} \to \mathcal{M}'$ is said to be equivariant (or, perhaps better, *$\rho$-equivariant*), if the diagram

$$\begin{array}{ccc} G \times \mathcal{M} & \xrightarrow{\rho \times f} & G' \times \mathcal{M}' \\ \downarrow{\scriptstyle \Gamma} & & \downarrow{\scriptstyle \Gamma'} \\ \mathcal{M} & \xrightarrow{f} & \mathcal{M}' \end{array} \tag{4.4.17}$$

commutes, so that

$$\Gamma' \circ \rho \times f = f \circ \Gamma \tag{4.4.18}$$

[8]Note that the idea of equivariance makes sense for a general group acting on a set: it is not necessary to use only Lie groups acting on manifolds.

or, equivalently,

$$\rho(g)f(p) = f(gp) \tag{4.4.19}$$

for all $g \in G$ and $p \in \mathcal{M}$.

An important physical application of this idea arises in the context of the theory of spinor fields on a general manifold $\mathcal{M}$. If $\mathcal{M}$ is spacetime, then the groups $G$ and $G'$ are $SL(2,\mathbb{C})$ and $SO(3,1)$ respectively; if $\mathcal{M}$ is physical space, the groups are $SU(2)$ and $SO(3)$ respectively. □

### 4.4.2 Different types of group action

We come now to a series of definitions concerned with various aspects of the action of a group $G$ on a set $\mathcal{M}$.

**Definition 4.9**

1. The *kernel* of a $G$-action is the subgroup of $G$ defined by

   $$K := \{g \in G \mid gp = p \text{ for all } p \in \mathcal{M}\}. \tag{4.4.20}$$

   The group action is *effective* if $K = \{e\}$.

2. The $G$-action is *free* if, for all $p \in \mathcal{M}$, $\{g \in G \mid gp = p\} = \{e\}$. Thus in a free action every point in $\mathcal{M}$ is moved away from itself by every element of $G$ with the exception of the unit element $e$ only.

   In particular, this means that given any pair of points $p, q \in \mathcal{M}$, either (i) there is no $g \in G$ such that $p = gq$; or (ii) there is a *unique* $g \in G$ such that $p = gq$.

3. The $G$-action is *transitive* if any pair of points $p, q \in \mathcal{M}$ can be 'connected' by an element of the group, *i.e.*, for all $p, q \in \mathcal{M}$ there exists $g \in G$ such that $p = gq$.

   In a transitive action the whole of $\mathcal{M}$ can be 'probed' by the $G$-action. This is complementary to the idea that in an effective action the whole of $G$ can be 'probed' by its action on $\mathcal{M}$.

4. Given that a generic $G$-action is not transitive, it is useful to define the *orbit* $O_p$ of the $G$-action through $p \in \mathcal{M}$ to be the set of all points in $\mathcal{M}$ that *can* be reached from $p$ (see Figure 4.1):

$$O_p := \{q \in \mathcal{M} \mid \exists g \in G \text{ with } q = gp\}. \tag{4.4.21}$$

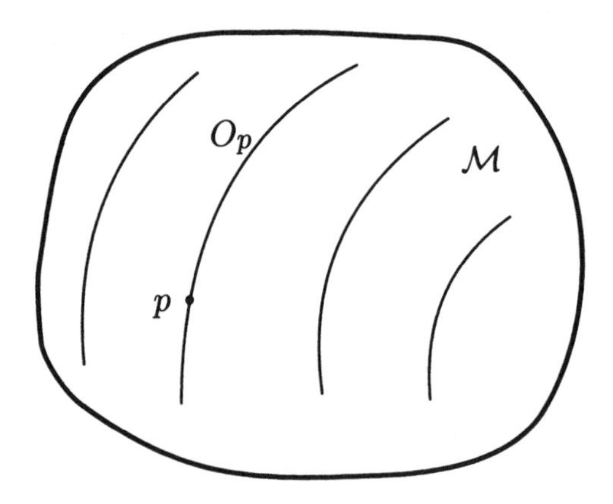


Figure 4.1: The orbits of a group action on $\mathcal{M}$.

5. If the $G$-action on $\mathcal{M}$ is transitive then the *stability group*, or *isotropy group*, or *little group*, $G_p$ of the action at a point $p \in \mathcal{M}$ is

$$G_p := \{g \in G \mid gp = p\}. \tag{4.4.22}$$

In the Lie case, this is a closed subgroup of $G$ for each $p \in \mathcal{M}$.

## Comments

1. The kernel measures the part of the group that is not involved at all in the $G$-action on $\mathcal{M}$. It is a normal subgroup of $G$ and the action of $G$ on $\mathcal{M}$ passes naturally to an action on $\mathcal{M}$ of the quotient group $G/K$ that *is* effective.

An example is afforded by the adjoint action of $G$ on itself in which $\mathrm{Ad}_g(g') := gg'g^{-1}$. The kernel of this action is the centre $C(G)$ of $G$. For example, in the adjoint action of $SU(3)$ on itself, the $\mathbf{Z}_3$ centre is represented trivially and it is the quotient group $SU(3)/\mathbf{Z}_3$

that acts effectively. The same remarks apply to the induced linear representation $\mathrm{Ad}_{g_*}$ on the vector space $T_eG$. Thus the eight-dimensional representation of $SU(3)$ (the 'octet' representation) is a faithful representation[9] of $SU(3)/\mathbf{Z}_3$. On the other hand, the three-dimensional ('quark') representation of $SU(3)$ is a faithful representation of $SU(3)$ itself.

2. A free action is effective but the converse is not necessarily true. For example, any faithful linear representation of $G$ is effective but it is never free since if $\vec{0}$ is the null vector in the representation vector space then, by linearity, $g\vec{0} = \vec{0}$ for all $g \in G$. Note that:

(i) A well-known example of a free action of any group $G$ is left (or right) translation on itself. This will play a central role in our discussion later of principal fibre bundles.

(ii) The adjoint action Eq. (4.2.20) of a group $G$ on itself is not free since $e \in G$ is left fixed by every $g \in G$, *i.e.*, $\mathrm{Ad}_g(e) = e$ for all $g \in G$.

3. To show that an action of a group $G$ on a set $\mathcal{M}$ is transitive, it suffices to show that there exists some point $p_0 \in \mathcal{M}$ with the property that, for any other $p \in \mathcal{M}$, there exists (at least one) $g \in G$ such that $p = gp_0$. Indeed, if this is the case, then given two points $p, q \in \mathcal{M}$ one can go from $p$ to $q$ by going first from $p$ to $p_0$ (using the inverse of the group element that transforms $p_0$ to $p$), and then from $p_0$ to $q$.

4. A very important example of a transitive $G$-action is when $\mathcal{M} = G/H$ where $H$ is a closed subgroup of $G$. The left action of $g \in G$ is defined by

$$\gamma_g(g'H) := (gg')H. \tag{4.4.23}$$

This action is transitive for any pair of groups $G$ and $H$, irrespective of whether they are Lie groups. However, in the case of Lie groups (with $H$ a closed subgroup) it can be shown that there is a unique

[9] A linear representation $g \mapsto L(g) \in \mathrm{Aut}(V)$ of a group $G$ on a vector space $V$ is said to be *faithful* if $\ker(L) = e$.

analytic manifold[10] structure on the coset space $G/H$ such that $G$ is a Lie transformation group on this space.

Note that:

(i) The action of $G$ on $G/H$ is effective if and only if $H$ contains no normal subgroup of $G$. It is never free since $h(eH) = eH$ for all $h \in H$.

(ii) A linear representation is never transitive since the null vector $\vec{0}$ cannot be taken to any other element in the vector space by the action of any $g \in G$.

5. It is clear that the orbits through any pair of points in $\mathcal{M}$ are either equal to each other or are disjoint (*cf.* Figure 4.1). This leads to an equivalence relation on $\mathcal{M}$ in which two points are defined to be equivalent if and only if they lie in the same orbit. As usual, $\mathcal{M}$ is partitioned into disjoint subsets (*i.e.*, the orbits) by this equivalence relation. An important physical example is the action of the gauge group on the space of Yang-Mills fields.

The set of equivalence classes is called the *orbit space* of the $G$-action on $\mathcal{M}$ and is denoted $\mathcal{M}/G$. If $\mathcal{M}$ is a topological space, then the orbit space carries a natural identification topology (see Eq. (1.4.45)), but it is frequently rather unpleasant (for example, non-Hausdorff).

Note that the orbit space of the right action of a subgroup $H$ of a group $G$ on $G$ is just the space of left cosets $G/H$. In the case of a Lie group, with $H$ a closed subgroup of $G$, the orbit space *does* have a decent topological/differential structure, namely the unique analytic structure mentioned above.

6. The kernel $K$ of a group action is related to the stability groups by

$$K = \bigcap_{p \in \mathcal{M}} G_p. \tag{4.4.24}$$

[10] A classic reference is Helgason (1962).

7. If $p$ and $q$ lie on the same orbit of a $G$-action on $\mathcal{M}$ then there is some $g \in G$ such that $p = gq$. It follows that

$$G_p = gG_q g^{-1} \tag{4.4.25}$$

so that the stability groups along an orbit are always conjugate (see Figure 4.2).

□

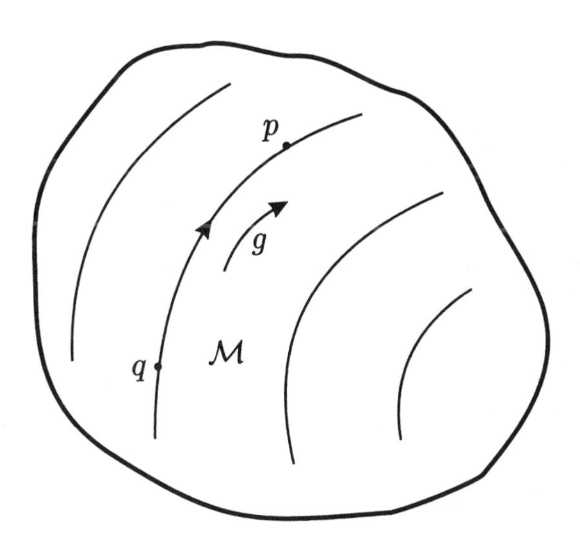


Figure 4.2: Stability groups along an orbit are conjugate.

### 4.4.3 The main theorem for transitive group actions

Next is a theorem of considerable significance, especially for the use of Lie groups in theoretical physics. As noted above, the action of $G$ on the coset space $G/H$ is transitive. The theorem of interest asserts the converse: if $\mathcal{M}$ is *any* space on which $G$ acts transitively then it is effectively of the form $G/H$ for some $H \subset G$. The word 'effectively' covers up a fair number of sins and it is best first to state the result from a purely group/set-theoretic perspective with no reference to topology or differential structure.

**Theorem 4.5** *Let $G$ be a group that acts transitively on a set $\mathcal{M}$. Then, for each $p \in \mathcal{M}$, there is a bijection $j_p : G/G_p \to \mathcal{M}$ defined by*

$$\begin{aligned} j_p : G/G_p &\to \mathcal{M} \\ gG_p &\mapsto gp \end{aligned} \tag{4.4.26}$$

*where $G_p$ is the stability group at the point $p$,*

## Proof

1. The first step is to ensure that the map in Eq. (4.4.26) is well-defined. Such a step is mandatory when—as is the case here—a function is defined on a space of equivalence classes in terms of a particular representative in each class: *i.e.*, it is necessary to show that the definition is independent of the particular representative that is selected.

In the present case, if $g_1$ and $g_2$ are in the same coset then $g_1 = g_2h$ for some $h \in G_p$, and hence $g_1p = g_2hp = g_2p$. Thus the definition in Eq. (4.4.26) is indeed independent of the representative chosen from the coset.

2. Suppose $j_p(g_1G_p) = j_p(g_2G_p)$. Then $g_1p = g_2p$ and hence $g_1^{-1}g_2p = p$, which means that $g_1^{-1}g_2$ belongs to the stability group $G_p$. This is precisely the condition that $g_1G_p = g_2G_p$ (*i.e.*, $g_1$ and $g_2$ belong to same right $G_p$-orbit) and so $j_p$ is one-to-one.

3. Since the $G$-action is transitive, every point in $\mathcal{M}$ can be written as $gp$ for some $g \in G$. Hence $j_p$ is surjective. **QED**

## Comments

1. When $G$ is a Lie group that acts smoothly and transitively on a differentiable manifold $\mathcal{M}$, a key question is whether or not the bijection $j_p$ in Eq. (4.4.26) is a diffeomorphism. The two major results in this direction are: (Chevalley 1946, Helgason 1962)

(i) The stability group $G_p$ is a *closed* subgroup of $G$, and hence $G/G_p$ can be equipped with the unique analytic manifold structure referred to earlier.

(ii) If $\mathcal{M}$ is locally compact and connected, and if $G$ is a locally compact Lie group, then the bijection $j_p$ is a diffeomorphism from $G/G_p$ onto $\mathcal{M}$.

2. If the $G$-action on $\mathcal{M}$ is *not* transitive then $\mathcal{M}$ decomposes into a disjoint union of orbits, on each of which $G$ *does* act transitively and with a specific stability group. It follows that each orbit $O_p$ is in bijective correspondence with the coset $G/G_p$ where $G_p$ is the stability group for $p \in O_p \subset \mathcal{M}$.

The critical question in the Lie group case is to decide when the spaces $G/G_p$ can be regarded as embedded submanifolds of $\mathcal{M}$, which first requires finding out when the isotropy groups of the different orbits are closed. This is a complicated question and for further information on these topological matters should be sought in the technical literature. □

### 4.4.4 Some important transitive actions

We shall now give several important examples of familiar manifolds that admit a transitive group of transformations by a Lie group $G$ and are hence diffeomorphic to a coset space $G/H$ for some subgroup $H$ of $G$.

**Theorem 4.6** *The $n$-sphere $S^n$ is diffeomorphic to the coset space $O(n+1,\mathbb{R})/O(n,\mathbb{R})$.*

**Proof**

1. According to the main theorem 4.5, it is necessary to find a transitive action of the group $O(n+1,\mathbb{R})$ on the $n$-sphere and then to show that the stability group at any conveniently chosen 'fiducial' point is a $O(n,\mathbb{R})$ subgroup.

To this end, let $\langle\vec{v},\vec{w}\rangle = \sum_{i=1}^{n+1} v_i w_i$ denote the usual inner product on the real vector space $\mathbb{R}^{n+1}$ where $\{v_i \mid i = 1,2,\ldots,n+1\}$ are the components of the vector $\vec{v}$ with respect to the usual orthonormal basis set $\{(1,0,\ldots,0),(0,1,\ldots,0),\ldots,(0,0,\ldots,1)\}$. Then the linear action of $O(n+1,\mathbb{R})$ on this vector space leaves invariant the inner product and hence, in particular, the set of unit-length vectors

$$S^n := \{\vec{v} \in \mathbb{R}^{n+1} \mid \langle \vec{v}, \vec{v} \rangle = 1\}. \tag{4.4.27}$$

This is the desired action of $O(n+1,\mathbb{R})$ on the $n$-sphere.

2. The first task is to show that this action is transitive. Let $\vec{v^0} := (1,0,\ldots,0) \in S^n \subset \mathbb{R}^{n+1}$, and let $\vec{v}$ be any other unit-length vector. Choose a set of orthonormal vectors $\{\vec{e_1}, \vec{e_2}, \ldots, \vec{e_n}\}$ such that the complete set $\{\vec{v}, \vec{e_1}, \vec{e_2}, \ldots, \vec{e_n}\}$ is an orthonormal basis set for $\mathbb{R}^{n+1}$. Then $\vec{v}$ satisfies the matrix equation $\vec{v} = A\vec{v^0}$, whose component form is

$$\begin{pmatrix} v_1 \\ v_2 \\ \vdots \\ v_{n+1} \end{pmatrix} = \begin{pmatrix} v_1 & e_1^1 & e_1^2 & \ldots & e_1^n \\ v_2 & e_2^1 & e_2^2 & \ldots & e_2^n \\ \vdots & \vdots & \vdots & \ddots & \vdots \\ v_{n+1} & e_{n+1}^1 & e_{n+1}^2 & \ldots & e_{n+1}^n \end{pmatrix} \begin{pmatrix} 1 \\ 0 \\ \vdots \\ 0 \end{pmatrix}. \tag{4.4.28}$$

However, the choice of the vectors $\{\vec{e_1}, \vec{e_2}, \ldots, \vec{e_n}\}$ is such that the matrix $A$ satisfies the equation $AA^t = A^tA = \mathbb{1}$ and hence $A \in O(n+1,\mathbb{R})$. Since $\vec{v}$ is any unit vector in the $n$-sphere, this shows that the action of $O(n+1,\mathbb{R})$ is transitive.

3. The isotropy group at the particular vector $\vec{v^0}$ is clearly all matrices of the form

$$\begin{pmatrix} 1 & 0 \; 0 \; \ldots \; 0 \\ 0 & \\ 0 & \\ \vdots & X \\ 0 & \end{pmatrix} \tag{4.4.29}$$

where $X$ is any $n \times n$ matrix that satisfies

$$XX^t = X^tX = \mathbb{1}_{n\times n}. \tag{4.4.30}$$

Hence the isotropy group is this particular[11] $O(n,\mathbb{R})$ subgroup of $O(n+1,\mathbb{R})$.

**QED**

[11] I say 'particular' subgroup, as there are many subgroups of $O(n+1,\mathbb{R})$ that are isomorphic to the subgroup of matrices satisfying Eq. (4.4.30); indeed any subgroup that is conjugate to this subgroup $H$ (*i.e.*, of the form $gHg^{-1}$ where $g \in O(n+1,\mathbb{R})$) has this property.

**Comments**

1. If the vectors $\{\vec{e_1}, \vec{e_2}, \ldots, \vec{e_n}\}$ are chosen such that the orientation on $\mathbb{R}^{n+1}$ induced by the orthonormal basis set $\{\vec{v}, \vec{e_1}, \vec{e_2}, \ldots, \vec{e_n}\}$ is the same as that arising from the usual basis, then the matrix $A$ in Eq. (4.4.28) satisfies $\det A = 1$ and hence belongs to the $SO(n+1, \mathbb{R})$ subgroup of $O(n+1, \mathbb{R})$. It then follows that

$$\boxed{S^n \cong SO(n+1, \mathbb{R})/SO(n, \mathbb{R}).} \tag{4.4.31}$$

2. A similar discussion applies to the invariance of $\langle \vec{v}, \vec{w} \rangle = \sum_{i=1}^{n+1} v_i^* w_i$ under the linear action of $U(n+1)$ on $\mathbb{C}^{n+1}$. This action will map the $(2n+1)$-sphere

$$S^{2n+1} \cong \{v \in \mathbb{C}^{n+1} \mid \langle v, v \rangle \equiv \sum_{i=1}^{n+1} |v_i|^2 = 1\} \tag{4.4.32}$$

into itself and—as in the case above—this action can be shown to be transitive. The stability group is a $U(n)$ subgroup, which gives the result

$$\boxed{S^{2n+1} \cong U(n+1)/U(n) \cong SU(n+1)/SU(n).} \tag{4.4.33}$$

A particular example is $n = 1$ when, since $SU(1) = \{e\}$, it follows that

$$S^3 \cong U(2)/U(1) \cong SU(2). \tag{4.4.34}$$

This result—that the group space of $SU(2)$ is diffeomorphic to a 3-sphere—is important in a number of different ways. It can also be obtained directly [exercise!] from a study of the implications of the defining relations

$$\begin{pmatrix} a & b \\ c & d \end{pmatrix} \begin{pmatrix} a^* & c^* \\ b^* & d^* \end{pmatrix} = \begin{pmatrix} 1 & 0 \\ 0 & 1 \end{pmatrix} \quad \text{and} \quad \det \begin{pmatrix} a & b \\ c & d \end{pmatrix} = 1 \tag{4.4.35}$$

for a matrix $\left(\begin{smallmatrix} a & b \\ c & d \end{smallmatrix}\right)$ to belong to $SU(2)$.

3. Another important example of a transitive group action can be obtained by considering the linear action of $SU(n+1)$ on $\mathbb{C}^{n+1}$. This clearly maps complex lines into complex lines and hence passes to an action on the complex projective space $\mathbb{C}P^n$ (which is defined to be the space of all such lines in $\mathbb{C}^{n+1}$). Unlike the original linear action, this new $SU(n+1)$ action *is* transitive, and the little group is a $U(n)$ subgroup. Hence:

$$\boxed{\mathbb{C}P^n \cong SU(n+1)/U(n).} \tag{4.4.36}$$

A particular example is $\mathbb{C}P^1 \cong SU(2)/U(1)$ which, since $\mathbb{C}P^1 \cong S^2$ (the complex Riemann sphere) proves the famous result

$$S^2 \cong SU(2)/U(1). \tag{4.4.37}$$

4. Similarly, the $SO(n+1,\mathbb{R})$ linear action on $\mathbb{R}^{n+1}$ generates the result

$$\boxed{\mathbb{R}P^n \cong SO(n+1,\mathbb{R})/O(n,\mathbb{R})} \tag{4.4.38}$$

where $\mathbb{R}P^n$ denotes the real projective space of all lines in $\mathbb{R}^{n+1}$. □

The next result is significant in the context of the group-theoretical structure of Riemannian geometry.

**Theorem 4.7** *Let $S_n$ denote the set of real, positive-definite, $n \times n$ symmetric matrices. Then*

$$S_n \cong GL^+(n,\mathbb{R})/SO(n,\mathbb{R}). \tag{4.4.39}$$

**Proof**

1. Define a left action of $GL^+(n,\mathbb{R})$ on $M(n,\mathbb{R})$ by

$$\gamma_g(A) := gAg^t \tag{4.4.40}$$

where $g \in GL^+(n,\mathbb{R})$ and $A \in M(n,\mathbb{R})$. The first step is to show that the subspace $S_n \subset M(n,\mathbb{R})$ is mapped into itself by this action.

(i) If $A = A^t$ then $(gAg^t)^t = (g^t)^t A^t g^t = gAg^t$, so that the image of a symmetric matrix is symmetric.

(ii) To say that $A$ is positive definite means that $\langle \vec{v}, A\vec{v} \rangle > 0$ for all $\vec{v} \in \mathbb{R}^n$ such that $\vec{v} \neq \vec{0}$, where $A$ is regarded as a linear operator on the real vector space $\mathbb{R}^n$ whose inner product $\langle \vec{v}, \vec{v} \rangle$ is defined as usual.

Then $\langle \vec{v}, (gAg^t)\vec{v} \rangle = \langle g^t\vec{v}, Ag^t\vec{v} \rangle$ since the transpose of a matrix is defined such that $\langle \vec{v}, B\vec{w} \rangle = \langle B^t\vec{v}, \vec{w} \rangle$. But $g \in GL^+(n, \mathbb{R})$ is invertible, and hence $\langle g^t\vec{v}, Ag^t\vec{v} \rangle > 0$ for all $\vec{v} \neq \vec{0}$ if and only if $\langle \vec{v}, A\vec{v} \rangle > 0$ for all $\vec{v} \neq \vec{0}$. Hence if $A$ is positive definite, so is $gAg^t$ for all $g \in GL^+(n, \mathbb{R})$.

These two statements show that $S_n$ is indeed mapped into itself under the $GL^+(n, \mathbb{R})$ action defined in Eq. (4.4.40).

2. To see that the action is transitive note that, since a symmetric matrix $S$ is diagonalisable, there exists some $O \in SO(n, \mathbb{R})$ and a diagonal matrix $D = \mathrm{diag}(d_1, d_2, \ldots, d_n)$ such that $S = ODO^t$. Furthermore, since $S$ is positive definite it follows that $d_i > 0$ for all $i = 1, 2, \ldots, n$, and hence the square root of $D$ can be taken to give

$$S = (OD^{\frac{1}{2}})\,\mathbb{1}\,(OD^{\frac{1}{2}})^t. \tag{4.4.41}$$

But Eq. (4.4.41) is of the form Eq. (4.4.40) and hence —since $OD^{\frac{1}{2}} \in GL^+(n, \mathbb{R})$—it follows that Eq. (4.4.41) is indeed a transformation of the type in Eq. (4.4.40). This shows that every $S \in S_n$ can be obtained from the unit matrix $\mathbb{1} \in S_n$ which is thus the 'fiducial' matrix for this action: in particular, the $GL^+(n, \mathbb{R})$ action on the subspace $S_n \subset M(n, \mathbb{R})$ is transitive.

3. The isotropy group at the fiducial matrix $\mathbb{1}$ is the set of all $g \in GL^+(n, \mathbb{R})$ such that $\gamma_g(\mathbb{1}) = g\,\mathbb{1}\,g^t = \mathbb{1}$. But this is precisely the subgroup $SO(n, \mathbb{R})$. This completes the proof of the theorem. **QED**

**Comments**

1. This result plays a central role in the fibre bundle approach to the theory of Riemannian metrics on a manifold where, roughly speaking, a Riemannian metric on a manifold $\mathcal{M}$ assigns an element of $S_n$ to each point in $\mathcal{M}$.

2. There is a similar result $S_{n,1} \cong GL^+(n, \mathbb{R})/SO(n-1, 1)$ where the matrices in $S_{n,1}$ have signature $(+, +, \cdots, +, -)$. In particular, the relation $S_{3,1} \cong GL^+(4, \mathbb{R})/SO(3, 1)$ is relevant in the mathematics of Lorentzian geometries in a four-dimensional spacetime. □

## 4.5 Infinitesimal Transformations

### 4.5.1 The induced vector field

The next topic is of considerable importance in any programme that uses Lie transformation groups: namely, the properties of 'infinitesimal transformations' and their relation to the full, globally-defined group action.

The basic idea is as follows (see Figure 4.3). If a Lie group $G$ acts as a group of transformations on a differentiable manifold $\mathcal{M}$ then each one-parameter subgroup of $G$ will produce a manifold-filling family of curves, and hence a vector field on $\mathcal{M}$ that is tangent to this family everywhere. By these means a map is obtained from the

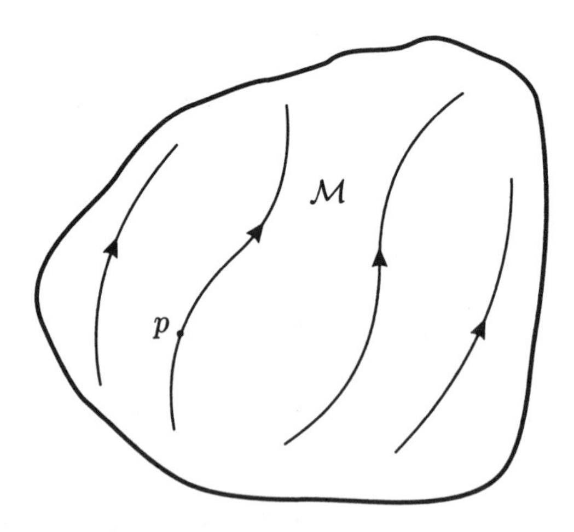


Figure 4.3: The orbits of a one-parameter subgroup of $G$.

Lie algebra element associated with the one-parameter subgroup into

the infinite-dimensional Lie algebra of all vector fields on $\mathcal{M}$. The interesting question is the extent to which this linear map reflects the details of the action of the Lie group $G$ on $\mathcal{M}$: as will be shown below, this is a generalisation of the way in which the Lie algebra of $G$ reflects the properties of $G$ itself.

In the discussion that follows a right-action of $G$ on $\mathcal{M}$ (rather than left-action) is used. The reason for this choice will emerge later, but it is not a significant restriction and the results obtained all have exact analogues for a left action.

**Definition 4.10**

Let $G$ be a Lie group that has a right action $g \to \delta_g$ on a differentiable manifold $\mathcal{M}$. Then the vector field $X^A$ on $\mathcal{M}$ *induced* by the action of the one-parameter subgroup $t \to \exp tA$, $A \in T_eG$, is defined as

$$X_p^A(f) := \frac{d}{dt} f(p \exp tA)\bigg|_{t=0} \tag{4.5.1}$$

where $f \in C^\infty(\mathcal{M})$, and $\delta_g(p)$ has been abbreviated to $pg$.

**Comments**

1. If the orbit $O_p$ through a point $p \in \mathcal{M}$ of the $G$-action is a smoothly embedded submanifold, then

$$T_p(O_p) \cong \{X_p^A \mid A \in T_eG\}. \tag{4.5.2}$$

2. The induced vector field $X^A$ has a flow $\phi_t^A(p) = p \exp tA = \delta_{\exp tA}(p)$; *i.e.*, the flow is

$$\phi_t^A = \delta_{\exp tA}. \tag{4.5.3}$$

3. The analogue of equations (4.4.10) and (4.4.11) for a right action is a smooth map $\Gamma : \mathcal{M} \times G \to \mathcal{M}$ with

(i) $\Gamma(p, e) = p$ for all $p \in \mathcal{M}$; (4.5.4)

(ii) $\Gamma(\Gamma(p, g_1), g_2) = \Gamma(p, g_1 g_2)$ for all $p \in \mathcal{M}$ and $g_1, g_2 \in G$. (4.5.5)

Then each $g \in G$ defines a map $\mathcal{M} \to \mathcal{M}$ by $\delta_g(p) := \Gamma(p, g) \equiv pg$.

However, there is another way of using $\Gamma : \mathcal{M} \times G \to \mathcal{M}$: namely, to each $p \in \mathcal{M}$ associate the map $\mathcal{M}_p : G \to \mathcal{M}$ defined by

$$\mathcal{M}_p(g) := \Gamma(p, g) \equiv pg \tag{4.5.6}$$

in terms of which Eq. (4.5.5) asserts the equivariance (see Eq. (4.4.15)) of $\mathcal{M}_p$ as

$$\mathcal{M}_p \circ r_g = \delta_g \circ \mathcal{M}_p \tag{4.5.7}$$

for all $g \in G$ and $p \in \mathcal{M}$ where $r_g : G \to G$ is the right translation $r_g(g') := g'g$.

Now, if $L^A$ is the left-invariant vector field on $G$ associated with $A \in T_eG$ then, for all $f \in C^\infty(\mathcal{M})$,

$$(\mathcal{M}_{p*} L^A_g)(f) = L^A_g(f \circ \mathcal{M}_p) = (l_{g*}A)(f \circ \mathcal{M}_p) = A(f \circ \mathcal{M}_p \circ l_g). \tag{4.5.8}$$

But

$$\mathcal{M}_p \circ l_g = \mathcal{M}_{pg} \tag{4.5.9}$$

for all $p \in \mathcal{M}$ and $g \in G$, and hence Eq. (4.5.8) implies

$$\begin{aligned} \left(\mathcal{M}_{p*} L^A_g\right)(f) = A(f \circ \mathcal{M}_{pg}) &= \frac{d}{dt} f \circ \mathcal{M}_{pg}(\exp tA)\Big|_{t=0} \\ &= \frac{d}{dt} f(pg \exp tA)\Big|_{t=0} . \end{aligned} \tag{4.5.10}$$

Thus, for all $g \in G$,

$$\mathcal{M}_{p*} L^A_g = X^A_{pg} \tag{4.5.11}$$

which shows that the vector fields $L^A$ on $G$ and $X^A$ on $\mathcal{M}$ are $\mathcal{M}_p$-related for each $p \in \mathcal{M}$.

Note that, since $L^A_e = A \in T_eG$, it follows from Eq. (4.5.11) that, for all $p \in \mathcal{M}$,

$$X^A_p = \mathcal{M}_{p*}(A) \tag{4.5.12}$$

which is a useful alternative definition of the induced field $X^A$. □

The expression Eq. (4.5.12) is used in the next result that deals with the infinitesimal version of the equivariance of a group action of the Lie group $G$ on a pair of manifolds $\mathcal{M}$ and $\mathcal{M}'$ (*cf.* Eq. (4.5.8)).

**Theorem 4.8** *Let $G$ be a Lie group that acts on the right on manifolds $\mathcal{M}$ and $\mathcal{M}'$ with induced vector fields $X^A$ and $X'^A$ respectively, and let $f : \mathcal{M} \to \mathcal{M}'$ be equivariant under this action;* i.e.,

$$f(pg) = f(p)g \tag{4.5.13}$$

*for all $p \in \mathcal{M}$ and $g \in G$. Then the vector fields $X^A$ and $X'^A$ are $f$-related for all $A \in T_eG$.*

## Proof

From Eq. (4.5.12) we have $X_p^A = \mathcal{M}_{p*}(A)$. Also, the equivariance condition Eq. (4.5.13) implies that, for all $g \in G$ and $p \in \mathcal{M}$,

$$(f \circ \mathcal{M}_p)(g) = f(pg) = f(p)g = \mathcal{M}'_{f(p)}(g) \tag{4.5.14}$$

and hence

$$f_*(X_p^A) = f_*\mathcal{M}_{p_*}(A) = (f \circ \mathcal{M}_p)_*(A) = \mathcal{M}'_{f(p)_*}(A) = X'^A_{f(p)}, \tag{4.5.15}$$

which is precisely the condition that the vector fields $X^A$ and $X'^A$ be $f$-related. **QED**.

Now let us consider a very special case of a right action of $G$ on a manifold $\mathcal{M}$: namely, when $\mathcal{M}$ is $G$ itself and $\delta_g = r_g$ for all $g \in G$. Then

$$X_g^A = G_{g_*}(A) = l_{g_*}(A) = L_g^A \tag{4.5.16}$$

since $G_g(g') = gg' = l_g(g')$. In other words, the left-invariant vector field $L^A$ is induced by the *right* translation of $G$ on itself. Thus, for all $f \in C^\infty(G)$,

$$L_g^A(f) = \frac{d}{dt} f(g \exp tA)\bigg|_{t=0} . \tag{4.5.17}$$

Similarly, the right-invariant vector field $R^A$, $A \in T_eG$, is induced from the *left*-translation operation as

$$R_g^A(f) = \frac{d}{dt} f((\exp tA)g)\bigg|_{t=0} . \tag{4.5.18}$$

This way of looking at the vector fields $L^A$ and $R^A$ is useful in a variety of situations. A good example is the following theorem which deals with the relation between the commutator of a pair of Lie algebra elements and the push-forward $\mathrm{Ad}_{g*}$ of the adjoint action of $G$ on itself.

**Theorem 4.9** *Let $A, B \in T_eG$ with Lie bracket $[AB]$. Then,*

$$[AB] = \frac{d}{dt}\mathrm{Ad}_{\exp tA_*}(B)\bigg|_{t=0}. \tag{4.5.19}$$

## Proof

In general, if $\phi_t^X$ is a flow for a vector field $X$ on a manifold $\mathcal{M}$ and if $Y$ is any other vector field then (see Eq. (3.2.20))

$$[X, Y] = -\frac{d}{dt}\phi_t^X{}_*(Y)\bigg|_{t=0} = \lim_{t\to 0}\frac{Y - \phi_t^X{}_*(Y)}{t}. \tag{4.5.20}$$

Now, according to Eq. (4.5.17), a flow for the left-invariant vector field $L^A$, $A \in T_eG$, on the Lie group $G$ is $\phi_t^A = r_{\exp tA}$. Thus, using Eq. (4.5.20),

$$\begin{aligned}
[AB] = [L^A, L^B]_e &= \lim_{t\to 0}\frac{L_e^B - r_{\exp tA_*}(L^B_{\exp -tA})}{t} \\
&= \lim_{t\to 0}\frac{B - r_{\exp tA_*}l_{\exp -tA_*}(B)}{t} \\
&= \lim_{t\to 0}\frac{B - \mathrm{Ad}_{\exp -tA_*}(B)}{t} \\
&= \lim_{t\to 0}\frac{\mathrm{Ad}_{\exp tA_*}(B) - B}{t}
\end{aligned} \tag{4.5.21}$$

which proves the result. **QED**

## Comments

1. We know from Eq. (4.2.44) that $\exp \mathrm{Ad}_{g*}(B) = g(\exp B)g^{-1}$ for all $g \in G$. Using this result and Eq. (4.5.19) it can be deduced that

$$\exp tA \exp B \exp -tA = \exp\left(t[AB] + O(t^2)\right). \tag{4.5.22}$$

□

The next, rather technical looking, theorem will be used in the proof of the main result in the next section.

**Theorem 4.10** *If the Lie group $G$ acts on a manifold $\mathcal{M}$ then, for all $g \in G$ and $A \in T_eG$,*

$$X^{\mathrm{Ad}_{g*}(A)} = \delta_{g^{-1}*}(X^A). \tag{4.5.23}$$

**Proof**

We have

$$X_p^{\mathrm{Ad}_{g*}(A)} = \mathcal{M}_{p*}\mathrm{Ad}_{g*}(A) = (\mathcal{M}_p \circ \mathrm{Ad}_g)_*(A). \tag{4.5.24}$$

But

$$\mathcal{M}_p \circ \mathrm{Ad}_g(g') = p(gg'g^{-1}) = \mathcal{M}_{pg} \circ r_{g^{-1}}(g') = \delta_{g^{-1}} \circ \mathcal{M}_{pg}(g') \tag{4.5.25}$$

so that $\mathcal{M}_p \circ \mathrm{Ad}_g = \delta_{g^{-1}}\mathcal{M}_{pg}$. Then Eq. (4.5.24) implies that

$$X_p^{\mathrm{Ad}_{g*}(A)} = \delta_{g^{-1}*}\mathcal{M}_{pg*}(A) = \delta_{g^{-1}*}(X_{pg}^A). \tag{4.5.26}$$

**QED**

### 4.5.2 The main result

We can now prove what is undoubtedly one of the central results in the theory of 'infinitesimal' transformations and which explains why the subject is important in so many applications. The theorem concerned asserts that the map that associates the vector field $X^A$ with $A \in T_eG$ is actually a *homomorphism* of Lie algebras: *i.e.*, the Lie algebra of the Lie group $G$ is 'represented' by the vector fields on the manifold $\mathcal{M}$ on which $G$ acts.

**Theorem 4.11** *Let $\mathcal{M}$ be a manifold on which a Lie group $G$ has a right action. Then the map $A \mapsto X^A$, which associates to each $A \in T_eG$ the induced vector field $X^A$ on $\mathcal{M}$, is a homomorphism of*

*$L(G) \cong T_eG$ into the infinite-dimensional Lie algebra of all vector fields on $\mathcal{M}$,* i.e.,

$$[X^A, X^B] = X^{[AB]} \tag{4.5.27}$$

*for all $A, B \in T_eG \cong L(G)$.*

**Proof**

It is clear from Eq. (4.5.12) that the map $A \mapsto X^A$ is linear, so the key task is to prove Eq. (4.5.27). Since (see Eq. (4.5.3)) a flow for $X^A$ is $\delta_{\exp(tA)}$, Eq. (4.5.20) can be used to write

$$[X^A, X^B] = \lim_{t\to 0} \frac{X^B - \delta_{\exp tA_*}(X^B)}{t}. \tag{4.5.28}$$

Hence, using Eq. (4.5.23),

$$\begin{aligned} [X^A, X^B] &= \lim_{t\to 0} \frac{X^B - X^{\mathrm{Ad}_{\exp -tA_*}(B)}}{t} \\ &= \lim_{t\to 0} \frac{X^{\mathrm{Ad}_{\exp tA_*}(B)-B}}{t} \end{aligned} \tag{4.5.29}$$

where the last step follows from the linearity properties $X^A - X^{A'} = X^{(A-A')}$ and $-X^A = X^{-A}$.

Since $T_eG$ is a finite-dimensional vector space, the linear map $A \mapsto X^A$ is continuous and hence the $\lim_{t\to 0}$ can be moved inside the square brackets to give

$$\begin{aligned} [X^A, X^B] &= X^{\lim_{t\to 0}(\mathrm{Ad}_{\exp tA_*}(B)-B)/t} \\ &= X^{[AB]} \end{aligned}$$

by virtue of Eq. (4.5.19). **QED**

**Comments**

1. If the same procedure had been followed throughout but with a left, rather than a right, action of $G$ on $\mathcal{M}$ then the result in Eq. (4.5.27) would have become

$$[X^A, X^B] = X^{[BA]} \tag{4.5.30}$$

so that the linear map $A \mapsto X^A$ would then be an *anti*-homomorphism from $T_eG \cong L(G)$ into the Lie algebra of vector fields on $\mathcal{M}$, rather than a homomorphism. This is the reason why a right action has been used throughout this section.

2. Several properties of the $G$-action on $\mathcal{M}$ are faithfully reflected in the homomorphism $\chi : T_eG \to \mathrm{VFld}(\mathcal{M})$, $\chi(A) := X^A$. For example:

(i) Suppose the action is effective and let $A \in \mathrm{Ker}(\chi)$, so that $\chi(A) = 0$. Then the one-parameter subgroup $t \mapsto \exp tA$ acts trivially on $\mathcal{M}$, which is consistent with the effective nature of the action only if $A = 0$. Hence we conclude:

"If $G$ acts effectively on $\mathcal{M}$ then the map $A \mapsto X^A$ is an *isomorphism* from $T_eG \cong L(G)$ into the Lie algebra of vector fields on $\mathcal{M}$."

(ii) Suppose the action is free and that $A \in T_eG$ is such that $X^A_p$ vanishes at some point $p \in \mathcal{M}$. This means that $p \in \mathcal{M}$ is left fixed by the one-parameter subgroup $t \mapsto \exp tA$, which contradicts the free nature of the action unless $A = 0$. Hence:

"If $G$ acts freely on $\mathcal{M}$ then the induced vector field $X^A$ is nowhere zero for all $A \in T_eG$, $A \neq 0$."

□

As a final remark on this whole issue of infinitesimal transformations one might wonder when the process discussed above can be reversed. That is, we have started with a group action of $G$ on $\mathcal{M}$ and derived a homomorphism from the Lie algebra of $G$ into the vector fields on $\mathcal{M}$. But if a manifold $\mathcal{M}$ and Lie group $G$ are such as to admit a homomorphism from $L(G)$ into the vector fields on $\mathcal{M}$, when is it true that there is a global group action of $G$ on $\mathcal{M}$ such that the given homomorphism is precisely the one that derives from this action?

One obvious problem is that usually more than one Lie group has a given Lie algebra, these being different global coverings of each other (for example, $SU(2)$ and $SO(3)$ have the same Lie algebra and there is a two-to-one covering homomorphism $\pi : SU(2) \to SO(3)$; see Eq. (5.1.8)). It is thus sensible to start with the assumption that the group concerned is the unique simply-connected Lie group with the given Lie algebra.

It is also necessary to restrict the discussion to those cases where the vector fields in the range of the homomorphism $\chi$ are complete since the fields induced by a group action always have this property. A famous result encompassing some of these features is *Palais' Theorem*.

"If $\chi$ is a homomorphism from $L(G)$ into the vector fields on $\mathcal{M}$, and if $G$ is simply connected and $\mathcal{M}$ is compact, then there exists a unique $G$-action on $\mathcal{M}$ such that $\chi(A) = X^A$—the induced vector field for that action."

# Chapter 5

# Fibre Bundles

## 5.1 Bundles in General

### 5.1.1 Introduction

In many situations in theoretical physics we encounter fields carrying indices. Some of these refer to tensorial properties with respect to the space on which the field is defined (often spacetime[1]); others are associated with internal structure on the space in which the field takes its values.

For example, consider the question of the precise mathematical structure that underlies the existence of an '$n$-tuple' of real scalar fields $\{\phi^1(x), \phi^2(x), \ldots, \phi^n(x)\}$ defined on a spacetime $\mathcal{M}$. One possibility is that it stems from a map $\phi : \mathcal{M} \to V$ from $\mathcal{M}$ to a real, $n$-dimensional vector space $V$. More precisely, if $\{\vec{e}_1, \vec{e}_2, \ldots, \vec{e}_n\}$ is some basis set for $V$, the value $\phi(x) \in V$ of the field at $x \in \mathcal{M}$ can

[1]The most common domain space for fields in modern physics is spacetime. For this reason I shall usually refer to the space on which a field is defined as 'spacetime', but it should be appreciated that this is not the only possibility. For example, in the canonical approach to a field theory the fields are defined on physical 3-space. And in special relativity, it is natural to consider fields whose domain is the 'mass-shell hyperbola' $p \cdot p = m^2$ in momentum space.

be expanded as

$$\phi(x) := \sum_{i=1}^{n} \phi^i(x)\vec{e}_i, \tag{5.1.1}$$

where, for all $x \in \mathcal{M}$, the quantities $\phi^i(x)$, $i = 1, 2, \ldots, n$, are real numbers. The coefficient functions $x \mapsto \phi^i(x)$, $i = 1, 2, \ldots, n$, are then the 'fields with indices'. In many cases of physical importance, the vector space $V$ arises as a representation space of some internal symmetry group.

Vector-valued fields are not the only possibility. Another important example is when $\phi : \mathcal{M} \to \mathcal{N}$ is a map from spacetime (or space) $\mathcal{M}$ to a manifold $\mathcal{N}$ that is *not* a vector space. For example, in the non-linear $\sigma$-model, the basic field takes its values in a homogeneous spaces $G/H$ where $H$ is a closed Lie subgroup of the Lie group $G$; as we saw in Chapter 2, such a space $G/H$ is typically a non-trivial manifold. In cases such as this, we are to think of having a local coordinate chart $(U, \psi)$ on $\mathcal{N}$ so that $\psi : U \subset \mathcal{N} \to \mathbb{R}^n$ with $\psi(y) := (\psi^1(y), \psi^2(y), \ldots, \psi^n(y)) \in \mathbb{R}^n$. The coefficients of our spacetime field $\phi$ are now defined by

$$\phi^i(x) := \psi^i(\phi(x)), \tag{5.1.2}$$

which makes sense provided the image point $\phi(x) \in \mathcal{N}$ lies in the open set $U \subset \mathcal{N}$. Note that vector-space valued fields are actually a special case of this situation in which we use the (global) coordinate chart on $V$ that is naturally associated with the basis set $\{\vec{e}_1, \vec{e}_2, \ldots, \vec{e}_n\}$ (*i.e.*, the coordinates of a vector $\vec{v} \in V$ are defined to be its components $v^i$ with respect to the basis set: $\vec{v} = \sum_{i=1}^{n} v^i \vec{e}_i$).

At first sight, the idea of a manifold-valued field $\phi : \mathcal{M} \to \mathcal{N}$ might appear to exhaust the possibilities of what is meant by a scalar 'field'. However, it has become clear that, in fact, this is not the case, and a genuine generalisation must be considered. This is the situation in which the space in which a field takes its values *varies* from point to point in the space $\mathcal{M}$ on which it is defined. Thus rather than writing $\phi : \mathcal{M} \to \mathcal{N}$ we must think now of a family of target spaces $\mathcal{N}_x$ labelled by the spacetime points $x \in \mathcal{M}$, and with $\phi(x) \in \mathcal{N}_x$ for each $x \in \mathcal{M}$. We are thus lead naturally to consider a bundle of spaces, parameterised by points in spacetime, in which our field appears as a '*cross-section*'.

Note that we have already encountered the idea of a bundle in Section 2.3 where the tangent spaces $T_x\mathcal{M}$ of a manifold $\mathcal{M}$ (parameterised by the points $x \in \mathcal{M}$) were 'glued' together to form the tangent bundle $T\mathcal{M}$. In particular, recall that a *vector field* on $\mathcal{M}$ can be defined as a map $X$ that associates to each point $x \in \mathcal{M}$ a point $X_x$ in the tangent space $T_x\mathcal{M}$; *i.e.,* it is a cross-section of the tangent bundle, and hence an example of a 'generalised field' of the type under consideration.

As we shall see later, tensor fields of all types appear naturally as cross-sections of various bundles associated with a spacetime manifold. Thus fibre bundle theory plays a central role in the underlying mathematics of differential geometry, and hence—in particular—of general relativity. However, this is not the only use of fibre bundle theory in physics. Another, very important, example arises in Yang-Mills theory in which the natural mathematical setting is a bundle of copies of the internal symmetry group $G$ (for example, $G$ might be $SU(2)$ or $SU(3)$), one at each point $x$ in spacetime. The way in which these fibres twist around the spacetime—typically Euclideanised and compactified—reflect topological properties of the gauge theory such as instanton number. A related application is the use of bundles in the context of Kaluza-Klein theories and the associated concept of 'dimensional reduction'.

Fibre bundle theory is an important part of pure mathematics too, especially algebraic topology. This is because the different ways in which the fibres of a bundle over a space $\mathcal{M}$ can twist around are determined by global topological properties of $\mathcal{M}$ that are specified by certain cohomology groups (see later). Indeed, in one powerful approach to algebraic topology ($K$-theory) the cohomology groups of a space $\mathcal{M}$ are essentially *defined* by the different bundles that can be built on it.

### 5.1.2 The definition of a bundle

Faced with a family of spaces $F_x$ parameterised by the elements $x \in \mathcal{M}$, it seems natural to require that the dependence of $F_x$ on the spacetime point $x$ be continuous, or even differentiable, in some way. For example, for the concrete case of a 2-sphere embedded in $\mathbb{R}^3$ it seems intuitively clear that the tangent spaces do indeed vary smoothly as we vary the point to which they are tangent, although it may not be obvious how this intuition should be made concrete in mathematical terms, especially if we do not wish to invoke the 'ambient' space $\mathbb{R}^3$ in which the 2-sphere appears as a hypersurface.

One possibility might be to think of $x \to F_x$ as a map from $\mathcal{M}$ to the collection $\mathcal{T}$ of all topological spaces. One could then require this map to be continuous with respect to some topology placed on $\mathcal{T}$, provided—of course—that a 'topologising' of $\mathcal{T}$ is possible. However, if $\mathcal{T}$ could be given a topology it would become a topological space, and hence a member of itself, thereby encountering Bertrand Russell's famous problem of barbers who shave themselves! But every set *can* be given at least one topology (for example, the discrete or indiscrete topology), and hence $\mathcal{T}$ is not a set but some more general class of objects.

A more productive approach is to consider the collection $E := \bigcup_{x\in\mathcal{M}} F_x$ of all the 'fibres' $F_x$, $x \in \mathcal{M}$, and require this to be a topological space in its own right. This idea leads to the basic mathematical definition of a bundle[2]:

**Definition 5.1**

1. A *bundle* is a triple $(E, \pi, \mathcal{M})$ where $E$ and $\mathcal{M}$ are topological spaces and $\pi : E \to \mathcal{M}$ is a continuous map.

2. The space $E$ is called the *bundle space*, or *total space*, of the bundle; $\mathcal{M}$ is the *base* space of the bundle; the map $\pi$ is the *projection*; and the inverse image $\pi^{-1}(\{x\})$ is the *fibre* over $x \in \mathcal{M}$.

The general idea of a bundle is sketched in Figure 5.1.

---

[2]Some of the best references on fibre bundles are also some of the oldest. A classic work—which makes heavy use of the idea of local bundle charts—is Steenrod (1951). A text that takes a more modern approach—and which has strongly influenced my own presentation—is Husemoller (1966)

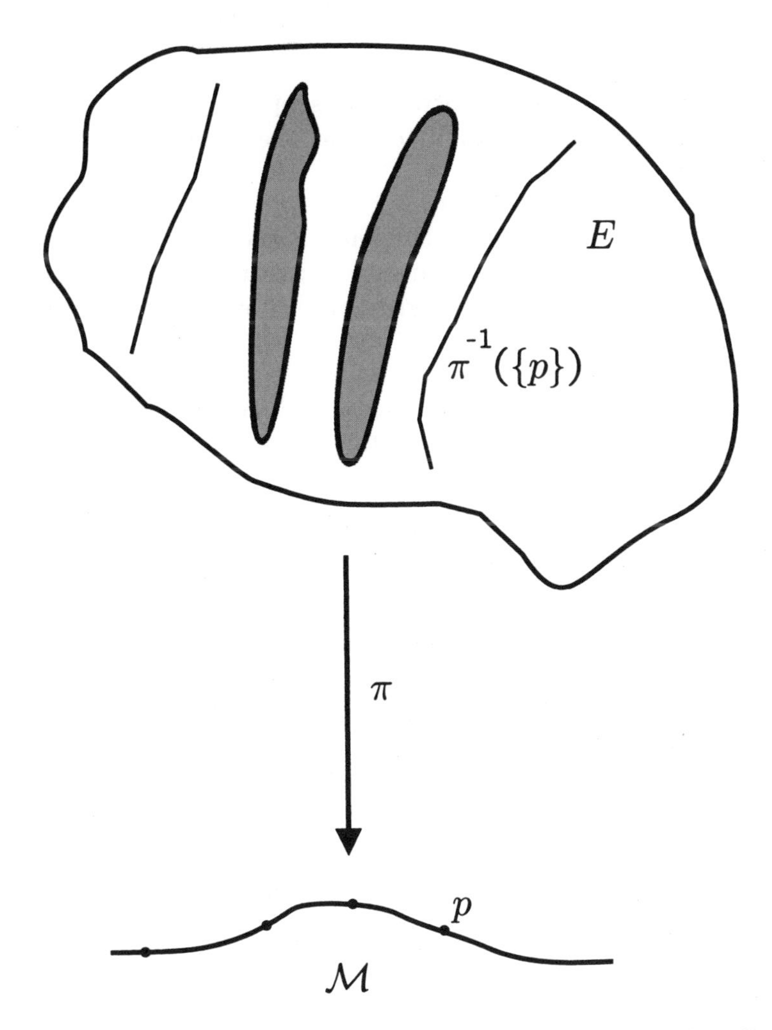


Figure 5.1: A diagrammatic representation of a general bundle.

**Comments**

1. There is no loss in generality in supposing that $\pi : E \to \mathcal{M}$ is *surjective*—if it is not, simply consider the bundle over the image $\pi(E) \subset \mathcal{M}$.

2. A $C^\infty$*-bundle* is as above but where $E$ and $\mathcal{M}$ are $C^\infty$-manifolds, and with $\pi$ a $C^\infty$-map. In what follows it is to be understood that any continuous map is also required to be $C^\infty$ if the spaces concerned are differentiable manifolds.

3. By common consent, the total space $E$ is often called the 'bundle' even though, strictly speaking, this refers to the triple $(E, \pi, \mathcal{M})$.

4. We often denote a bundle (*i.e.,* the triple) with a Greek letter like $\xi$ or $\eta$. In this case, the total and base spaces of a bundle $\xi$ are denoted $E(\xi)$ and $\mathcal{M}(\xi)$ respectively.

5. The definition above of a bundle is very general. However, in all existing applications in physics (and for most uses in pure mathematics too) the bundles that arise have the special property that the fibres $\pi^{-1}(\{x\})$, $x \in \mathcal{M}$, are all homeomorphic (diffeomorphic, in the manifold case) to a *common* space $F$. In this situation, $F$ is known as the *fibre* of the bundle, and the bundle is said to be a *fibre bundle*. This condition will always apply in the cases of interest to us, and from now on we shall mainly assume that it is satisfied.

If $(E, \pi, \mathcal{M})$ is a bundle with fibre $F$ it is often convenient to indicate this by writing $F \to E \xrightarrow{\pi} \mathcal{M}$ or, sometimes,

$$\begin{array}{ccc} F & \longrightarrow & E \\ & & \downarrow{\scriptstyle \pi} \\ & & \mathcal{M} \end{array} \tag{5.1.3}$$

depending on what is typographically convenient. □

**Examples**

1. One of the simplest examples of a fibre bundle is the *product bundle* over $\mathcal{M}$ with fibre $F$. This is defined simply as the triple $(\mathcal{M} \times F, \mathrm{pr}_1, \mathcal{M})$. It is sketched in Figure 5.2.

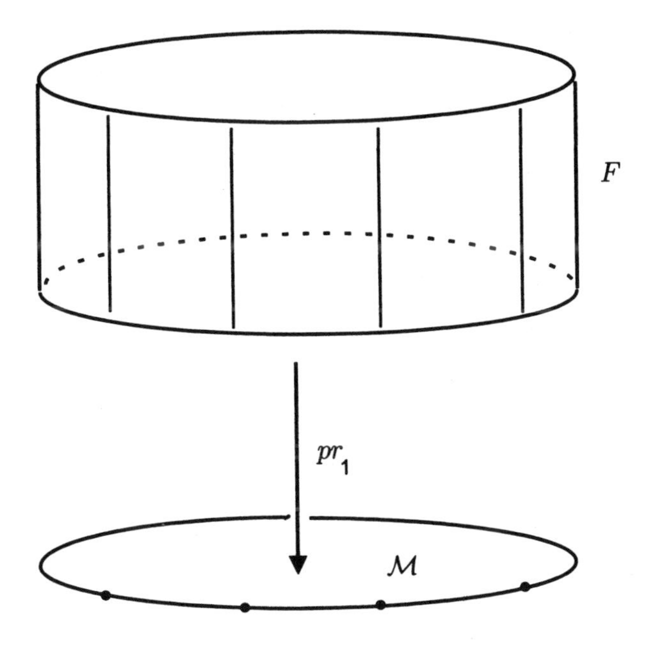


Figure 5.2: A product bundle.

2. A famous example of a fibre bundle is the *Möbius band* which is a twisted strip whose base space is the circle $S^1$. Thus, for example, the fibre could be taken to be the closed interval $[-1, 1]$. However, note that the total space $E$ is *not* the product space $S^1 \times [-1, 1]$ (and neither is it homeomorphic to it). It can be represented by taking a rectangle and identifying the short edges, as shown in Figure 5.3—in particular, the two points '$b$' (resp. '$c$') on the two sides of the strip are to be identified. Note that, in this example, the circle is represented by the line with the two endpoints '$a$' identified.

3. Another famous example is the *Klein bottle*. This is rather difficult to draw since it cannot be embedded in Euclidean 3-space. However, it can be represented with the aid of the diagram in Figure 5.4 which shows how it is obtained from a cylinder by reflecting in the diameter $d$–$e$ and then identifying points.

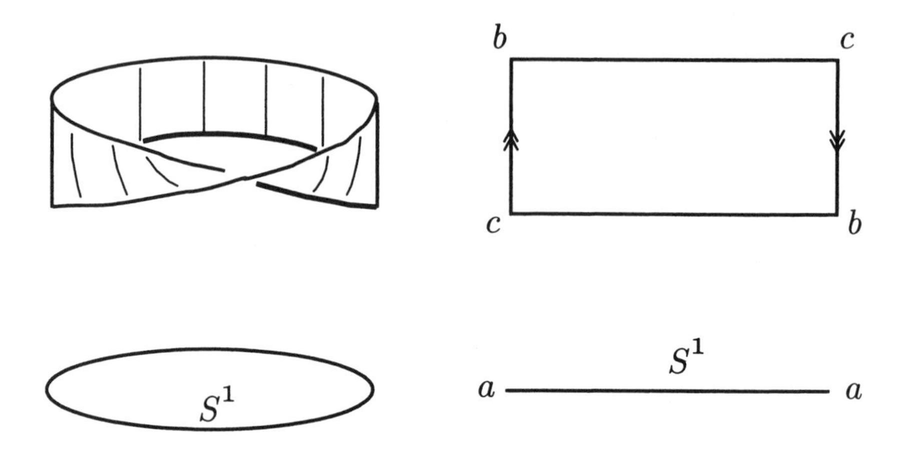


Figure 5.3: The Möbius strip.

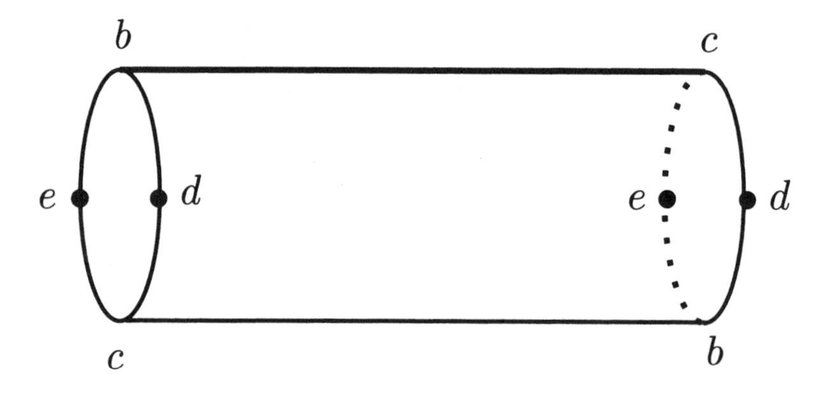


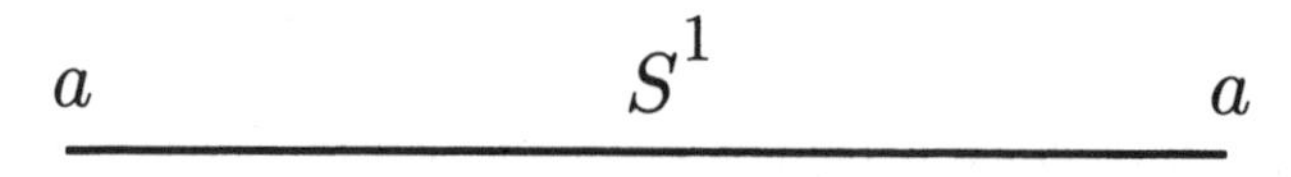


Figure 5.4: The Klein bottle.

### 5.1.3 The idea of a cross-section

As mentioned already, from the viewpoint of applications to theoretical physics, one of the most important ideas in fibre bundle theory is that of a 'cross-section'. In the earlier chapters on differential geometry, we saw that vector fields and one-forms can be regarded as cross-sections of the tangent bundle and cotangent bundle respectively, and we shall see later how this idea can be extended to incorporate general tensor fields. In Yang-Mills theory, all fields (with the exception of the Yang-Mills field itself) are cross-sections of various vector bundles that are associated with the fundamental Yang-Mills 'principal' fibre bundle of parameterised copies $G_x$, $x \in \mathcal{M}$, of the internal symmetry group $G$. Cross-sections of vector bundles always exist, but this is not necessarily the case for a general fibre bundle; an important example of this phenomenon will be given below in the context of principal bundles.

**Definition 5.2**

A *cross-section* of a bundle $(E, \pi, \mathcal{M})$ is a map $s : \mathcal{M} \to E$ such that the image of each point $x \in \mathcal{M}$ lies in the fibre $\pi^{-1}(\{x\})$ over $x$. More precisely,

$$\pi \circ s = \mathrm{id}_{\mathcal{M}}. \tag{5.1.4}$$

**Comments**

1. In the case of a product fibre bundle $(\mathcal{M} \times F, \mathrm{pr}_1, \mathcal{M})$, a cross-section $s$ defines a unique function $\tilde{s} : \mathcal{M} \to F$ given by

$$s(x) = (x, \tilde{s}(x)) \tag{5.1.5}$$

for all $x \in \mathcal{M}$; conversely, each such function $\tilde{s}$ gives a unique cross-section $s$ defined as Eq. (5.1.5). This is sketched in Figure 5.5. Thus, in a product bundle, a cross-section is equivalent to a normal function from the base space $\mathcal{M}$ to the fibre $F$.

2. The situation for a non-product fibre bundle $F \to E \xrightarrow{\pi} \mathcal{M}$ is different: a cross-section of a bundle of this type can at best be

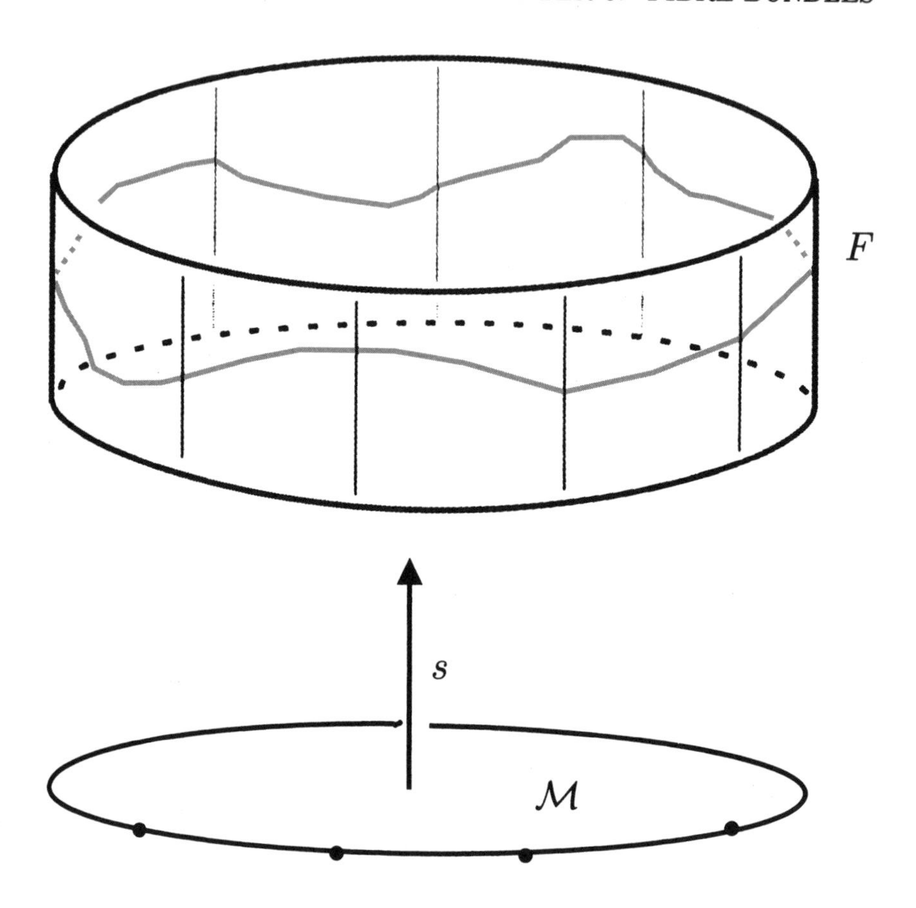


Figure 5.5: A cross-section of a product bundle.

regarded as a type of 'twisted' function from $\mathcal{M}$ to $E$. For example, in the case of a Möbius band, a cross-section can be represented using the identification diagram in Figure 5.6. This shows that a cross-section of the Möbius bundle is equivalent to a function from $S^1$ to $[-1, 1]$ that is *anti*-periodic around the circle.

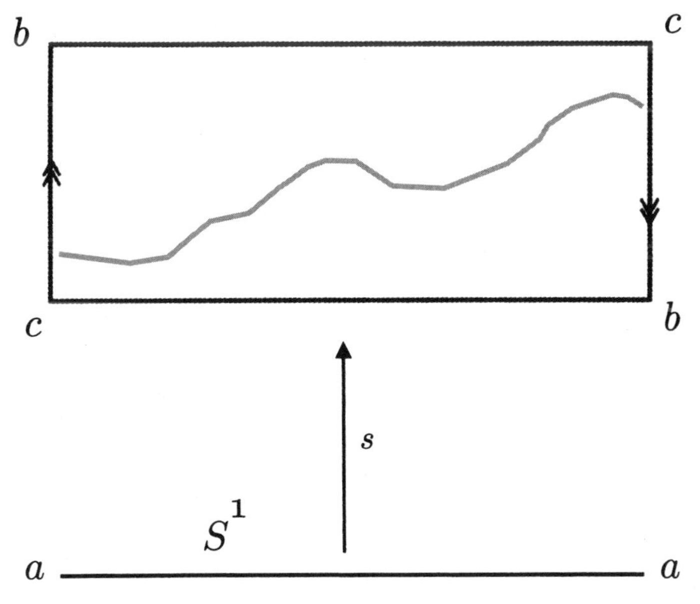


Figure 5.6: A cross-section of a Möbius bundle.

**Examples**

1. Let $H$ be a closed Lie subgroup of the Lie group $G$. Define the map $\pi : G \to G/H$ by $\pi(g) := gH$. Then $(G, \pi, G/H)$ is a bundle with fibre $H$. In general there will be no (smooth) cross-sections for fibre bundles of this type since, as will become clear below, a necessary condition for the existence of such sections is that $G$ is diffeomorphic to $G/H \times H$, which is usually not true.

2. The tangent bundle $TS^n$ of the $n$-sphere $S^n$ can be represented as a sub-bundle of the product bundle $(S^n \times \mathbb{R}^{n+1}, \mathrm{pr}_1, S^n)$ with a bundle space defined as

$$E(TS^n) := \{(\vec{x}, \vec{y}) \in S^n \times \mathbb{R}^{n+1} \mid \vec{x} \cdot \vec{y} = 0\} \tag{5.1.6}$$

where the $n$-sphere $S^n$ is identified as the surface $\{\vec{x} \in \mathbb{R}^{n+1} \mid \vec{x} \cdot \vec{x} = 1\}$ in $\mathbb{R}^{n+1}$.

Similarly, the *normal-bundle* $\nu(S^n)$ of $S^n$ embedded in $\mathbb{R}^{n+1}$ is defined to be the set of all vectors in $\mathbb{R}^{n+1}$ that are normal to points on the sphere:

$$E(\nu(S^n)) := \{(\vec{x}, \vec{y}) \in S^n \times \mathbb{R}^{n+1} \mid \exists r \in \mathbb{R} \text{ such that } \vec{y} = r\vec{x}\}. \tag{5.1.7}$$

Note that, with this definition, a cross-section of $TS^n$ is clearly equivalent to a vector field on $S^n$. A cross-section of $\nu(S^n)$ is called a *normal field* of $S^n$ as a submanifold of $\mathbb{R}^{n+1}$.

### 5.1.4 Covering spaces and sheaves

A covering space is a special type of bundle that arises naturally in a number of places in theoretical physics. The idea of a covering is also a useful introduction to the theory of sheaves, which plays a crucial role in some fundamental areas of pure mathematics, especially topos theory.

**Definition 5.3**

A topological space $\tilde{\mathcal{M}}$ *covers* another space $\mathcal{M}$ if there is a continuous map $\pi : \tilde{\mathcal{M}} \to \mathcal{M}$ with the property that each $x \in \mathcal{M}$ has an open neighbourhood $U \subset \mathcal{M}$ such that $\pi^{-1}(U)$ is a disjoint union of open sets $U_i$, each of which is mapped homeomorphically onto $U$ by the map $\pi$.

**Examples**

1. A simple example of a covering space is the double covering of a circle by itself under the map $\pi : S^1 \to S^1$ defined by $\pi(e^{i\theta}) := e^{2i\theta}$, where $S^1$ is represented by the set of all complex numbers of modulus one. This is sketched in Figure 5.7.

2. Another example of a covering of the circle is the map $\pi : \mathbb{R} \to S^1$ defined by $\pi(r) := e^{ir}$ in which the covering space is the real line.

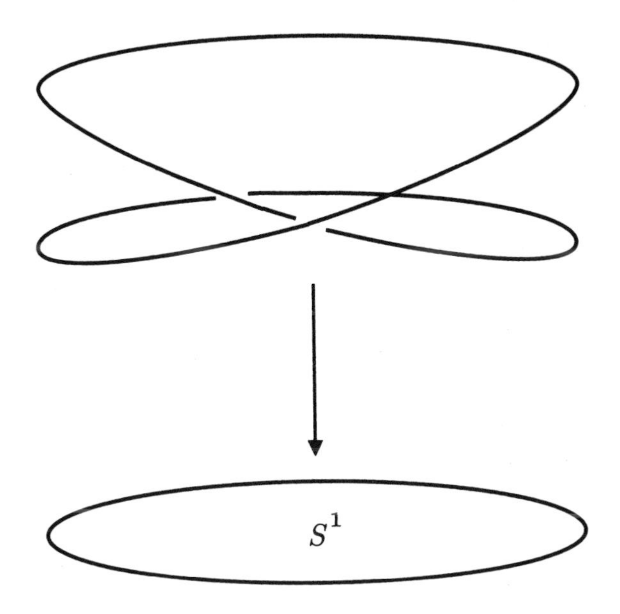


Figure 5.7: A double covering of a circle by a circle.

3. A more sophisticated example is the map $\pi : SU(2) \to SO(3)$ which takes a $2 \times 2$ complex matrix $A \in SU(2)$ to the orthogonal matrix

$$\pi(A)_{ij} := \frac{1}{2}\mathrm{tr}\,(A\tau_i A^{-1}\tau_j) \tag{5.1.8}$$

where $\tau_i$, $i = 1, 2, 3$, are the Pauli spin-matrices. Viewed group theoretically, this map is a group homomorphism with kernel

$$\mathbf{Z}_2 \simeq \left\{ \begin{pmatrix} 1 & 0 \\ 0 & 1 \end{pmatrix}, \begin{pmatrix} -1 & 0 \\ 0 & -1 \end{pmatrix} \right\}. \tag{5.1.9}$$

Viewed topologically, it defines a two-to-one covering of $\mathbb{R}P^3$ (the group manifold of $SO(3)$) by $S^3$ (the group manifold of $SU(2)$).

Running parallel to the theory of bundles is the theory of 'sheaves'. Sheaves play a fundamental role in many branches of pure mathematics (for example: complex function theory, algebraic geometry, generalised cohomology theory, the theory of topoi and the logical foundations of mathematics) and—although they have not yet appeared much in theoretical physics—it is worth giving a few of the basic definitions.[3]

**Definition 5.4**

1. A bundle $(E, \pi, \mathcal{M})$ is said to be *étale* if the projection map $\pi : E \to \mathcal{M}$ is a local homeomorphism—*i.e.*, to each $p \in E$ there exists some open set $U \subset E$ with $p \in U$ and such that (i) $\pi(U)$ is an open subset of $\mathcal{M}$; and (ii) $\pi_{|U}$ is a homeomorphism of $U$ with $\pi(U)$.

2. A *sheaf of abelian groups* is an étale bundle $(E, \pi, \mathcal{M})$ in which each fibre $\pi^{-1}(\{x\})$, $x \in \mathcal{M}$, is an abelian group.

   There are analogous definitions for a *sheaf of rings* and other such algebraic structures.

**Comments**

1. The product bundle $(\mathcal{M} \times F, \mathrm{pr}_1, \mathcal{M})$ can never be étale if $F$ is any connected $C^\infty$ manifold. Thus, in some sense, étale bundles are complementary to our intuitive idea of a twisted product of manifolds.

2. A classic example of an étale bundle is a covering space. However, not every étale map is a covering. Thus covering maps are local homeomorphisms in a rather strong sense.

3. There is an intimate connection between the idea of a sheaf and that of a 'presheaf' on a space $\mathcal{M}$. The latter is an assignment of an abelian group (or ring *etc*) $A(U)$ to each open subset $U \subset \mathcal{M}$, and an assignment, to each pair $U \subset V$ of open sets in $\mathcal{M}$, of a homomorphism $\rho_{UV} : A(V) \to A(U)$ (called the *restriction map*) such that (i) $\rho_{UU} = 1$; and (ii) if $U \subset V \subset W$ then

$$\rho_{UV}\,\rho_{VW} = \rho_{UW}. \tag{5.1.10}$$

A well-known example is the assignment to each open subset $U$ of the ring $C^\infty(U)$ of $C^\infty$-functions on $U$. In this case, the 'restriction

[3]A comprehensive and sophisticated reference to sheaves in the context of topos theory is MacLane & Moerdijk (1992).

map' is precisely what the name implies: *i.e.*, if $U \subset V$, the function $\rho_{UV} : C^\infty(V) \to C^\infty(U)$ is defined to take a $C^\infty$-function on $V$ to its restriction to the subset $U$.

Every sheaf $(\Lambda, \pi, \mathcal{M})$ gives rise naturally to an associated presheaf —namely the presheaf of local cross-sections[4] of $\Lambda$. Conversely, from each presheaf one can construct an associated sheaf of 'germs'. Presheaves and sheaves play an important role in algebraic topology where, for example, singular cochains can take their values in locally specified abelian groups.

### 5.1.5 The definition of a sub-bundle

Earlier, when discussing the general ideas of differentiable manifolds, it was pointed out that many important examples arise naturally as sub-manifolds of other, more tractable, manifolds. A similar situation arises for fibre bundles and leads to the idea of a 'sub-bundle' of a bundle. The formal definition is as follows.

**Definition 5.5**

1. A bundle $(E', \pi', \mathcal{M}')$ is a sub-bundle of a bundle $(E, \pi, \mathcal{M})$ if
   (a) $E' \subset E$;
   (b) $\mathcal{M}' \subset \mathcal{M}$;
   (c) $\pi' = \pi_{|E'}$.
2. Let $\mathcal{N}$ be a subspace of $\mathcal{M}$. Then the *restriction* of $(E, \pi, \mathcal{M})$ to $\mathcal{N}$ is defined to be the bundle $(\pi^{-1}(\mathcal{N}), \pi_{|_{\pi^{-1}(\mathcal{N})}}, \mathcal{N})$. Note that if $(E, \pi, \mathcal{M})$ has fibre $F$ then so does its restriction to $\mathcal{N} \subset \mathcal{M}$.

We recall that cross-sections of a product bundle $(\mathcal{M} \times F, \mathrm{pr}_1, \mathcal{M})$ are in one-to-one correspondence with functions $\tilde{s} : \mathcal{M} \to F$ according to Eq. (5.1.5). More generally, if $(E, \pi, \mathcal{M})$ is a sub-bundle of the

[4]A *local cross-section* of a bundle is a cross-section of the bundle restricted to some open subset of the base space. The definition of a 'restriction' of a bundle is given in Definition 5.5.

product bundle $(\mathcal{M} \times F, \mathrm{pr}_1, \mathcal{M})$ then cross-sections of the former have the form

$$s(x) = (x, \tilde{s}(x)) \tag{5.1.11}$$

where $\tilde{s} : \mathcal{M} \to F$ is any function such that, for all $x \in \mathcal{M}$, $(x, \tilde{s}(x)) \in E$.

### 5.1.6 Maps between bundles

Now we come to the important idea of a 'bundle map'[5] between a pair of bundles $(E, \pi, \mathcal{M})$ and $(E', \pi', \mathcal{M}')$. This involves continuous maps from $E$ to $E'$ and $\mathcal{M}$ to $\mathcal{M}'$ which map fibres over $\mathcal{M}$ into fibres over $\mathcal{M}'$ (as usual, these maps are required to be smooth if $E$, $E'$, $\mathcal{M}$, and $\mathcal{M}'$ are differentiable manifolds). More precisely:

**Definition 5.6**

A *bundle map* between a pair of bundles $(E, \pi, \mathcal{M})$ and $(E', \pi', \mathcal{M}')$ is a pair of maps $(u, f)$ where $u : E \to E'$ and $f : \mathcal{M} \to \mathcal{M}'$ such that the following diagram is commutative

$$\begin{array}{ccc} E & \xrightarrow{u} & E' \\ \downarrow{\scriptstyle \pi} & & \downarrow{\scriptstyle \pi'} \\ \mathcal{M} & \xrightarrow{f} & \mathcal{M}' \end{array} \tag{5.1.12}$$

*i.e.*,

$$\pi' \circ u = f \circ \pi \tag{5.1.13}$$

**Comments**

1. The equation (5.1.13) implies precisely that, for all $x \in \mathcal{M}$, $u(\pi^{-1}(\{x\})) \subset \pi'^{-1}(\{f(x)\})$, *i.e.*, the pair of maps $(u, f)$ maps fibres into fibres.

2. Since $\pi$ is surjective, the map $f$ is uniquely determined by the map $u$.

---

[5]More precisely, this is a *morphism* in the category of bundles—*i.e.*, a structure-preserving map between the relevant objects in the category.

3. A special example of a bundle map is an *isomorphism* between a pair of bundles $(E, \pi, \mathcal{M})$ and $(E', \pi', \mathcal{M}')$. This is defined to be a bundle map $(u, f)$ from $(E, \pi, \mathcal{M})$ to $(E', \pi', \mathcal{M}')$ for which there is another bundle map $(u', f')$ from $(E', \pi', \mathcal{M}')$ to $(E, \pi, \mathcal{M})$ such that

$$u' \circ u = \mathrm{id}_E; \quad u \circ u' = \mathrm{id}_{E'}; \tag{5.1.14}$$

$$f' \circ f = \mathrm{id}_{\mathcal{M}}; \quad f \circ f' = \mathrm{id}_{\mathcal{M}'}. \tag{5.1.15}$$

4. Given maps $f : \mathcal{M} \to \mathcal{M}'$ and $f' : \mathcal{M}' \to \mathcal{M}''$, the composition $f' \circ f : \mathcal{M} \to \mathcal{M}''$ is defined by $f' \circ f(x) := f'(f(x))$ for all $x \in \mathcal{M}$. Similarly, the *composition* of the bundle map $(u, f) : (E, \pi, \mathcal{M}) \to (E', \pi', \mathcal{M}')$ and the bundle map $(u', f') : (E', \pi', \mathcal{M}') \to (E'', \pi'', \mathcal{M}'')$ is the bundle map between the bundles $E$ and $E''$ defined as $(u' \circ u, f' \circ f) : (E, \pi, \mathcal{M}) \to (E'', \pi'', \mathcal{M}'')$.

The following commutative diagram helps to illustrate this:

$$\begin{array}{ccccc} E & \xrightarrow{u} & E' & \xrightarrow{u'} & E'' \\ \downarrow{\scriptstyle \pi} & & \downarrow{\scriptstyle \pi'} & & \downarrow{\scriptstyle \pi''} \\ \mathcal{M} & \xrightarrow{f} & \mathcal{M}' & \xrightarrow{f'} & \mathcal{M}'' \end{array} \tag{5.1.16}$$

5. If $\mathcal{M} = \mathcal{M}'$, a bundle map $(u, \mathrm{id}_{\mathcal{M}})$ is called a $\mathcal{M}$-*bundle map*. Note that in this case each fiber in the bundle is mapped into itself. □

Note that fibre bundles are usually thought of as being some sort of 'twisted product' of the base space with the fibre—a rather pictorial concept which implies that, in some appropriate sense, a fibre bundle looks like a product 'in the small'. This restriction is not contained in the definition of a bundle given above (although, in fact, all the examples given are of this type) and the first task is to define what it means for two bundles to look like each other locally; we can then require a general fibre bundle to look locally like a product bundle.

**Definition 5.7**

1. Two bundles $\xi$ and $\eta$ with the same base space $\mathcal{M}$ are said to be *locally isomorphic* if, for each $x \in \mathcal{M}$, there exists an open neighbourhood $U$ of $x$ such that $\xi_{|U}$ and $\eta_{|U}$ are $U$-isomorphic.

2. A fibre bundle $(E, \pi, \mathcal{M})$ is *trivial* if it is $\mathcal{M}$-isomorphic to the product bundle $(\mathcal{M} \times F, \mathrm{pr}_1, \mathcal{M})$ for some space $F$.

   It is *locally trivial* if it is locally isomorphic to the product bundle.

Note that

(a) The relation of being locally isomorphic is an equivalence relation on the set of all bundles over the topological space (or differentiable manifold) $\mathcal{M}$.

(b) When a bundle is trivial there is not in general any distinguished, or uniquely determined, trivialisation.

**Examples**

The bundles in the previous examples are all locally trivial. Note that a locally trivial bundle is necessarily one with a fibre. In fact, an alternative *definition* of a fibre bundle with fibre $F$ is a triple $(E, \pi, \mathcal{M})$ such that, for each $x \in \mathcal{M}$, there exists an open neighbourhood $U \subset \mathcal{M}$ of $x$ and a homeomorphism $h : U \times F \to \pi^{-1}(U)$ such that

$$\pi(h(x, y)) = x \text{ for all } x \in U,\ y \in F. \tag{5.1.17}$$

From now on, all bundles under discussion will be assumed to be locally trivial.

### 5.1.7 The pull-back operation

We come now to an idea of the greatest importance. Recall that one of the most significant features of differential forms (when compared with vector fields) is that, given a pair of manifolds $\mathcal{M}$ and $\mathcal{N}$ and a map $f : \mathcal{M} \to \mathcal{N}$, any differential form $\omega$ on $\mathcal{N}$ can be 'pulled back' to give a differential form $f^*(\omega)$ on $\mathcal{M}$. It is highly significant that a fibre bundle can also be pulled back by a map into its base space. In particular, this shows the deep consistency between bundle theory in general and the specific use of differential forms on the base space $\mathcal{M}$ of a bundle to describe the twists of the fibres around the base space. As might be expected, a central role is played here by the cohomology groups of $\mathcal{M}$ and their relation to differential forms via DeRham's theorem.

**Definition 5.8**

Let $\xi = (E, \pi, \mathcal{M})$ be a bundle with base space $\mathcal{M}$, and let $f : \mathcal{M}' \to \mathcal{M}$ be any map from $\mathcal{M}'$ to $\mathcal{M}$. Then the *induced bundle* (or *pull-back*) of $\xi$ is the bundle $f^*(\xi)$ over $\mathcal{M}'$ with a total space

$$E' := \{(x', e) \in \mathcal{M}' \times E \mid f(x') = \pi(e)\} \tag{5.1.18}$$

and with the projection map $\pi' : E' \to \mathcal{M}'$ defined by $\pi'(x', e) := x'$.

**Comments**

1. There is a natural bundle map from the induced bundle $(E', \pi', \mathcal{M}')$ to the original bundle $(E, \pi, \mathcal{M})$, defined as:

$$\begin{array}{ccc} E(f^*(\xi)) & \xrightarrow{f_\xi} & E(\xi) \\ \downarrow{\scriptstyle \pi'} & & \downarrow{\scriptstyle \pi} \\ \mathcal{M}' & \xrightarrow{f} & \mathcal{M} \end{array} \tag{5.1.19}$$

where $f_\xi(x', e) := e$.

Note that $\pi'^{-1}(\{x'\}) = \{(x', e) \in \{x'\} \times E \mid f(x') = \pi(e)\} = \{x'\} \times \pi^{-1}(\{f(x')\})$, and hence $f_\xi$ maps each fibre $\pi'^{-1}(\{x'\}) \subset E(f^*(\xi))$ homeomorphically onto the fibre $\pi^{-1}(\{f(x')\}) \subset E(\xi)$. In particular, this shows that each fibre of the pull-back $f^*(\xi)$ is homeomorphic to the fibre $F$ of the original bundle $\xi$.

2. If $j : \mathcal{N} \to \mathcal{M}$ is the inclusion map for a subspace $\mathcal{N}$ of $\mathcal{M}$ then the restriction $\xi_{|\mathcal{N}}$ is $\mathcal{N}$-isomorphic to the induced bundle $j^*(\xi)$.

3. If $g : \mathcal{M}'' \to \mathcal{M}'$ and $f : \mathcal{M}' \to \mathcal{M}$, then if $\xi = (E, \pi, \mathcal{M})$ is a bundle over $\mathcal{M}$, the pull-backs obey the same type of functorial property as do the pull-backs of differential forms. More precisely, the bundles $g^*(f^*(\xi))$ and $(f \circ g)^*(\xi)$ are $\mathcal{M}''$-isomorphic.

4. If $f : \mathcal{M}' \to \mathcal{M}$ is a smooth map between manifolds $\mathcal{M}'$ and $\mathcal{M}$ then the functions on $\mathcal{M}$ can be pulled back by $f$ to give functions on $\mathcal{M}'$. Specifically, if $\phi \in C^\infty(\mathcal{M})$, define $f^*(\phi) \in C^\infty(\mathcal{M}')$ by $f^*(\phi)(x) := \phi(f(x))$ for all $x \in \mathcal{M}'$. It is noteworthy that cross-sections of bundles pull back too. Thus if $s : \mathcal{M} \to E(\xi)$ is a cross-section of $\xi$, an associated cross-section $\tilde{s} : \mathcal{M}' \to E(f^*(\xi))$ can be defined by

$$\tilde{s}(x) := (x, s(f(x))). \tag{5.1.20}$$

In studying this definition it is helpful to look at the following diagram

$$\begin{array}{ccccccc} & E(f^*(\xi)) & \xrightarrow{f_\xi} & E(\xi) & \\ \uparrow \tilde{s} & \downarrow \pi' & & \downarrow \pi & \uparrow s \\ & \mathcal{M}' & \xrightarrow{f} & \mathcal{M} & \end{array} \tag{5.1.21}$$

which commutes in the sense that $f_\xi \circ \tilde{s} = s \circ f$.

### 5.1.8 Universal bundles

There is a deep algebraic-topological connection between the classification of bundles over some fixed base space and the pull-back operation. This starts with the observation that if $\xi = F \to E \to \mathcal{N}$ is a fibre bundle over $\mathcal{N}$, and if $f$ is a map from $\mathcal{M}$ into $\mathcal{N}$, then it seems intuitively clear that the pull-back bundle $f^*(\xi)$ will not change dramatically if $f$ is varied in a 'continuous' way. More precisely, if $f, g$ are a pair of maps from $\mathcal{M}$ into $\mathcal{N}$ that are *homotopic* then it is fairly easy to show that $f^*(\xi)$ and $g^*(\xi)$ are isomorphic bundles over $\mathcal{M}$.

Thus, if $\mathcal{B}_F(\mathcal{M})$ denotes the set of isomorphism classes of $F$-bundles over $\mathcal{M}$, any $F$-bundle $\xi$ over $\mathcal{N}$ generates a map

$$\begin{aligned} \alpha_\xi : [\mathcal{M}, \mathcal{N}] &\to \mathcal{B}_F(\mathcal{M}) \\ f &\mapsto f^*(\xi) \end{aligned} \tag{5.1.22}$$

where $[\mathcal{M}, \mathcal{N}]$ denotes the set of homotopy classes of maps[6] from $\mathcal{M}$ into $\mathcal{N}$.

[6]Strictly speaking, these maps have to be base-point preserving, *i.e.*, we choose points $x_0 \in \mathcal{M}$, $y_0 \in \mathcal{N}$ and require that all maps $f$ from $\mathcal{M}$ to $\mathcal{N}$ under consideration satisfy $f(x_0) = y_0$.

In general, the map in Eq. (5.1.22) is neither one-to-one nor onto. However, for a wide class of spaces $\mathcal{M}$ (for example CW-complexes, which includes all differentiable manifolds) there exist *universal bundles* $\mathcal{U}F = F \to EF \to BF$ with the property that the map

$$\alpha_{\mathcal{U}F} : [\mathcal{M}, BF] \to \mathcal{B}_F(\mathcal{M}) \tag{5.1.23}$$

is both one-to-one and onto. Thus:

1. if $f, g : \mathcal{M} \to BF$ then $f^*(\mathcal{U}F) \simeq g^*(\mathcal{U}F)$ if, and only if, $f$ and $g$ are homotopic maps;

2. if $\eta$ is any $F$-bundle over $\mathcal{M}$, there exists some $f : \mathcal{M} \to BF$ such that $\eta$ is $\mathcal{M}$-isomorphic to $f^*(\mathcal{U}F)$.

It can be shown that *any* $F$-bundle whose total space $EF$ is *contractible* can serve as a model for a universal $F$-bundle.[7] We shall see some concrete examples of this in the next section on principal bundles.

Thus the set of isomorphism classes of $F$-bundles over $\mathcal{M}$ is in bijective correspondence with the set of homotopy classes of maps from $\mathcal{M}$ into $BF$. This procedure is rendered meaningful by the observation that different universal $F$-bundles have base spaces that are necessarily homotopically equivalent. Hence the problem of classifying bundles over $\mathcal{M}$ 'reduces' to the problem of computing the homotopy set $[\mathcal{M}, BF]$—a task that is tailor-made for study with the aid of algebraic topology.

The general idea of 'characteristic classes' can also be seen clearly from this construction. Thus let $\omega$ be an element of some cohomology group of $BF$ and let $\eta$ be a bundle over a space $\mathcal{M}$ induced by a map $\phi : \mathcal{M} \to BF$, *i.e.*, $\eta \simeq \phi^*(\mathcal{U}F)$. Then the *characteristic class* of $\eta$ corresponding to the 'universal' class $\omega$ is defined as

$$\omega_\eta := \phi^*(\omega). \tag{5.1.24}$$

Note that this object is unambiguously defined since any pair of maps $\phi_1$ and $\phi_2$ from $\mathcal{M}$ into $BF$ that induce isomorphic bundles over $\mathcal{M}$

[7] For example, see Husemoller (1966), or Milnor & Stasheff (1974).

are homotopic, and hence the induced cohomology classes $\phi_1^*(\omega)$ and $\phi_2^*(\omega)$ are equal.

Note also that if $\psi : \mathcal{M} \to \mathcal{N}$ and $\phi : \mathcal{N} \to BF$ with $\eta = \phi^*(\mathcal{U}F)$, then $(\phi \circ \psi)^*(\mathcal{U}F) \simeq \psi^*(\phi^*(\mathcal{U}F) \simeq \psi^*(\eta)$, and hence

$$\begin{aligned}\psi^*(\omega_\eta) &= \psi^*(\phi^*(\omega)) = (\phi \circ \psi)^*(\omega) = \omega_{(\phi\circ\psi)^*(\mathcal{U}F)} = \omega_{\psi^*(\phi^*(\mathcal{U}F))} \\ &= \omega_{\psi^*(\eta)}. \end{aligned} \tag{5.1.25}$$

Thus the association of a characteristic class $\omega_\eta$ with a bundle $\eta$ is 'covariant' with respect to the pull-back operation of bundles and cohomology classes. In particular, if a model can be found for $BF$ that is a differentiable manifold, then one can define real characteristic classes of an $F$-bundle on a manifold $\mathcal{M}$ as the pull-backs of DeRham cohomology classes on $BF$.

## 5.2 Principal Fibre Bundles

### 5.2.1 The main definition

We start now to consider the wide-ranging question of how fibre bundles are to be classified, and how the different types can be constructed. A crucial idea is that of a 'principal fibre bundle', which is a bundle whose fibre is a Lie group in a certain special way. These have the important property that all non-principal bundles are associated with an underlying principal bundle. Furthermore, the twists in a bundle associated with a particular principal bundle are uniquely determined by the twists in the latter, and hence the topological implications of fibre-bundle theory are essentially coded into the theory of principal fibre bundles. The crucial definitions are as follows.

**Definition 5.9**

1. A bundle $(E, \pi, \mathcal{M})$ is a *G-bundle* if $E$ is a right $G$-space and if $(E, \pi, \mathcal{M})$ is isomorphic to the bundle $(E, \rho, E/G)$ where $E/G$ is the orbit space of the $G$-action on $E$ and $\rho$ is the usual projection

map:

$$\begin{array}{ccc} E & \xrightarrow{u} & E \\ \downarrow{\scriptstyle \pi} & & \downarrow{\scriptstyle \rho} \\ \mathcal{M} & \xrightarrow{\simeq} & E/G \end{array} \tag{5.2.1}$$

Note that the fibres of the bundle are the orbits of the $G$-action on $E$, and hence—in general—will not be homeomorphic to each other.

2. If $G$ acts *freely* on $E$ then $(E, \pi, \mathcal{M})$ is said to be a *principal G-bundle*, and $G$ is then called the *structure group* of the bundle. The freedom of the $G$-action implies that each orbit is homeomorphic to $G$, and hence we have a fibre bundle with fibre $G$.

**Examples**

1. The product bundle $(\mathcal{M} \times G, \mathrm{pr}_1, \mathcal{M})$ is a principal $G$-bundle under the obvious right $G$-action $(x, g_0)g := (x, g_0 g)$. One might expect a 'trivial' $G$-bundle to be defined as one that is isomorphic to the product bundle. However, this requires a definition of a 'morphism' between two principal bundles, and this is *not* the same thing as a bundle map since the $G$-action must also be preserved; we shall return to this later.

2. Let $G$ be the cyclic group $\mathbf{Z}_2 = \{e, a\}$ with $a^2 = e$, and let the non-identity element of this group act on the $n$-sphere $S^n$ by exchanging the antipodal points: $\vec{x}e := \vec{x}$, $\vec{x}a := -\vec{x}$; here $S^n$ is viewed as the set of unit-length vectors $\vec{x}$ in $\mathbb{R}^{n+1}$. This $\mathbf{Z}_2$-action is free and gives rise to a principal $\mathbf{Z}_2$-bundle whose base space is diffeomorphic to the real projective space $\mathbb{R}P^n \simeq S^n/\mathbf{Z}_2$. A special case is the bundle $\mathbf{Z}_2 \to S^3 \to \mathbb{R}P^3$ discussed earlier in the context of the covering map Eq. (5.1.8).

3. One of the most important examples of a principal bundle occurs when $H$ is a closed Lie subgroup of a Lie group $G$. Then $H$ acts on the right on $G$ by group multiplication with an action that is clearly free and whose orbit space is simply the space of cosets $G/H$. Thus we get a principal $H$-bundle $(G, \pi, G/H)$ whose fibre is

the group $H$. When this bundle is locally trivial (see below) the Lie group $G$ can be thought of globally as being the 'twisted' product of $H$ with $G/H$.

For example, the $U1$ action on $SU(2)$ gives rise to a bundle $U1 \to SU(2) \to SU(2)/U1$, which is $S^1 \to S^3 \to S^2$. This famous fibering of the three-sphere by the circle is known as the *Hopf* bundle.

Other examples of principal bundles obtained in this way are

$$U(n) \to SU(n+1) \to \mathbb{C}P^n \tag{5.2.2}$$

and

$$SO(n) \to SO(n+1) \to S^n \tag{5.2.3}$$

and

$$SU(n) \to SU(n+1) \to S^{2n+1}. \tag{5.2.4}$$

4. In Yang-Mills theory with the internal symmetry group $SU(2)$, the bundle corresponding to instanton number one is a principal $SU(2)$-bundle whose base space is a four-sphere regarded as the one-point compactification of 'Euclideanised' Minkowski spacetime. We recall that the group space of $SU(2)$ is a three-sphere, and in fact this instanton bundle is of the form $S^3 \to S^7 \to S^4$. This is clearly not a product bundle since $S^7$ is not homeomorphic to $S^3 \times S^4$.

5. Another example of a principal bundle can be obtained by considering the action of the multiplicative group $\mathbb{C}_*$ of non-zero complex numbers on the space $\mathbb{C}^{n+1} - \{0\}$ given by

$$(z_1, z_2, \ldots, z_{n+1})\lambda := (z_1\lambda, z_2\lambda, \ldots, z_{n+1}\lambda) \tag{5.2.5}$$

for all $\lambda \in \mathbb{C}_*$.

This is a free action and the orbit space is the space $\mathbb{C}P^n$ of complex lines in the vector space $\mathbb{C}^{n+1}$. Thus we get the principal bundle

$$\mathbb{C}_* \to \mathbb{C}^{n+1} - \{0\} \to \mathbb{C}P^n \tag{5.2.6}$$

where the projection map associates with each $n+1$-tuple

$$(z_1, z_2, \ldots, z_{n+1})$$

the point in $\mathbb{C}P^n$ whose homogeneous coordinates are $(z_1; z_2; \ldots; z_{n+1})$.

A closely connected bundle is obtained by restricting to the $U1$ subgroup of $\mathbb{C}_*$ which acts freely on the real $(2n+1)$-sphere embedded in $\mathbb{C}^{n+1} - \{0\}$. This gives rise to the principal $U1$-bundle

$$U1 \to S^{2n+1} \to \mathbb{C}P^n. \tag{5.2.7}$$

6. A principal fibre bundle of major importance in differential geometry is the bundle of frames attached to an $m$-dimensional differentiable manifold $\mathcal{M}$. A *linear frame* (or *base*) $b$ at a point $x \in \mathcal{M}$ is an ordered set $(b_1, b_2, \ldots, b_m)$ of basis vectors for the tangent space $T_x\mathcal{M}$. The total space $\mathbf{B}(\mathcal{M})$ of the bundle of frames is defined to be the set of all frames at all points in $\mathcal{M}$, and the projection map $\pi : \mathbf{B}(\mathcal{M}) \to \mathcal{M}$ is defined to be the function that takes a frame into the point in $\mathcal{M}$ to which it is attached. A natural, free right action of $GL(m, \mathbb{R})$ on $\mathbf{B}(\mathcal{M})$ is defined by

$$(b_1, b_2, \ldots, b_m)g := \Big(\sum_{j_1=1}^{m} b_{j_1} g_{j_1 1}, \sum_{j_2=1}^{m} b_{j_2} g_{j_2 2}, \ldots \sum_{j_m=1}^{m} b_{j_m} g_{j_m m}\Big) \tag{5.2.8}$$

for all $g \in GL(m, \mathbb{R})$.

The bundle space $\mathbf{B}(\mathcal{M})$ can be given the structure of a differentiable manifold as follows. Let $U \subset \mathcal{M}$ be a coordinate neighbourhood with coordinate functions $(x_1, x_2, \ldots, x_m)$. Then any base $b = (b_1, b_2, \ldots, b_m)$ for the vector space $T_x\mathcal{M}$, $x \in U$, can be expanded uniquely as

$$b_i = \sum_{j=1}^{m} b_i^j \left(\frac{\partial}{\partial x^j}\right)_x, \quad i = 1, 2, \ldots, m \tag{5.2.9}$$

for some non-singular matrix $b_i^j \in GL(m, \mathbb{R})$. Thus we can define a map

$$\begin{aligned} h : U \times GL(m, \mathbb{R}) &\to \pi^{-1}(U) \\ (x, g) &\mapsto \Big(\sum_{j_1=1}^{m} g_1^{j_1} (\partial_{j_1})_x, \sum_{j_2=1}^{m} g_2^{j_2} (\partial_{j_2})_x, \ldots, \sum_{j_m=1}^{m} g_m^{j_m} (\partial_{j_m})_x\Big) \end{aligned} \tag{5.2.10}$$

and then use $(x^1, x^2, \ldots, x^m; g_i^j)$ as the coordinates for a differential structure on $\mathbf{B}(\mathcal{M})$. By these means, $\mathbf{B}(\mathcal{M})$ becomes a manifold of dimension $m + m^2$.

7. Another important bundle in Yang-Mills theory is obtained by considering the set $\mathcal{A}$ of all Yang-Mills potentials in the theory. If the underlying $G$-bundle over spacetime $\mathcal{M}$ is trivial, then the corresponding gauge group $\mathcal{G}$ is the set of all[8] smooth maps from $\mathcal{M}$ into $G$; more generally it is the group of automorphisms of the principal bundle (see below). In either case, the physical configurations of the theory are identified with potentials that are related by gauge transformations, *i.e.*, they lie in the same orbit of the action of the gauge group $\mathcal{G}$ on $\mathcal{A}$. With some care in the precise selection of $\mathcal{G}$ and/or $\mathcal{A}$ it can be shown that this action is free and hence that $\mathcal{A}$ is a principal $\mathcal{G}$-bundle with base space $\mathcal{A}/\mathcal{G}$ (Singer 1978).

A related example arises in general relativity with the space Riem$\mathcal{M}$ of Lorentzian (or Riemannian, if that is appropriate) metrics on $\mathcal{M}$ on which the diffeomorphism group of $\mathcal{M}$ acts as a type of analogue of the gauge group $\mathcal{G}$ of Yang-Mills theory with $g \in \text{Riem}\mathcal{M}$ being taken to $\phi^* g$ under the action of $\phi \in \text{Diff}\mathcal{M}$. Both these examples differ from the previous ones in that the spaces concerned are infinite-dimensional, and hence some subtlety arises in the detailed constructions.

### 5.2.2 Principal bundle maps

After this fairly extensive set of examples we must return to the general theory and, in particular, to the generalisation of the idea of a bundle map to include the preservation of the group action in a principal bundle.

**Definition 5.10**

1. A bundle map $(u, f)$ between a pair of principal $G$-bundles $(P, \pi, \mathcal{M})$ and $(P', \pi', \mathcal{M}')$ is said to be a *principal* bundle map

[8] More precisely, a gauge transformation is required to have *compact* support, *i.e.*, it is equal to the unit element of $G$ for all spacetime points that lie outside some compact subset of $\mathcal{M}$. It is important to make this distinction in order to separate genuine gauge transformations from 'rigid' (*i.e.*, spacetime independent) transformations that may appear as symmetries of the Hamiltonian or action.

(or just a *principal map*) if $u : P \to P'$ is $G$-equivariant in the sense that

$$u(pg) = u(p)g \tag{5.2.11}$$

for all $p \in P$ and $g \in G$.

2. This can be generalised to the case in which $(P, \pi, \mathcal{M})$ is a $G$-bundle, $(P', \pi', \mathcal{M}')$ is a $G'$-bundle, and $\Lambda : G \to G'$ is a group homomorphism. We then require

$$u(pg) = u(p)\Lambda(g) \tag{5.2.12}$$

for all $p \in P$, and $g \in G$.

An example of the generalised situation arises in the theory of spinor fields on a Riemannian manifold $\mathcal{M}$ (see also the remarks following Eq. (4.4.17)). In this case, $G'$ is the orthogonal group $SO(m, \mathbb{R})$ (where $m = \dim \mathcal{M}$) that acts on the bundle of orthonormal frames, and $G$ is its double covering group—denoted $\mathrm{Spin}(m, \mathbb{R})$, and called the *spin group* of $\mathcal{M}$—that acts on the bundle of spinor frames.[9] In the case of a Lorentzian manifold, $G'$ is the Lorentz group $SO(m-1, 1; \mathbb{R})$. In the special case of $SO(3,1)$, the corresponding spinor group $\mathrm{Spin}(3,1;\mathbb{R})$ happens to be isomorphic to $SL(2, \mathbb{C})$.

We come now to a theorem that is of some significance, although it is easy enough to prove.

**Theorem 5.1** *Let $(u, \mathrm{id}_{\mathcal{M}})$ be a principal $\mathcal{M}$-map between a pair of principal $G$-bundles $(P, \pi, \mathcal{M})$ and $(P', \pi', \mathcal{M})$:*

$$\begin{array}{ccc} P & \xrightarrow{u} & P' \\ \downarrow{\scriptstyle \pi} & & \downarrow{\scriptstyle \pi'} \\ \mathcal{M} & \xrightarrow{\mathrm{id}_{\mathcal{M}}} & \mathcal{M} \end{array} \tag{5.2.13}$$

*Then $u$ is necessarily an* isomorphism.

**Proof**

(i) First we shall show that $u$ is one-to-one. Thus let $p_1 \in P$ and $p_2 \in P$ be such that $u(p_1) = u(p_2)$. Then $\pi(p_1) = \pi'(u(p_1)) =$

[9] For a full discussion of the spinor group $\mathrm{Spin}(m, \mathbb{R})$, see Husemoller (1966).

$\pi'(u(p_2)) = \pi(p_2)$, and therefore $p_1$ and $p_2$ belong to the same fibre. Hence there exists $g \in G$ such that $p_1 = p_2 g$, and thus $u(p_1) = u(p_2 g) = u(p_2)g = u(p_1)g$. But $G$ acts freely on $P$, and hence $g = e$. Thus $p_1 = p_2$, which proves that $u$ is one-to-one.

(ii) Next let us show that $u$ is surjective. Thus let $p' \in P'$, and choose $p \in P$ such that $\pi(p) = \pi'(p')$. Then $\pi'(u(p)) = \pi(p) = \pi'(p')$ and hence $p'$ and $u(p)$ are in the same fibre. Therefore there exists $g \in G$ such that $p' = u(p)g = u(pg)$, and so $u$ is surjective.

(iii) Finally, it must be shown that the inverse map $u^{-1}$ is continuous (or smooth, in the case of differentiable manifolds). This is left as an exercise.[10] **QED**

### Comments

1. Armed with the idea of a principal bundle map we can now give a proper definition of what it means to say that a principal $G$-bundle $(P, \pi, \mathcal{M})$ is *trivial*. Namely, that there is a principal bundle map from $(P, \pi, \mathcal{M})$ to the product bundle $(\mathcal{M} \times G, \mathrm{pr}_1, \mathcal{M})$.

The idea of *local triviality* is then defined in the obvious way as the requirement that $(P, \pi, \mathcal{M})$ is locally trivial as a bundle, and for each open set $U \subset \mathcal{M}$ on which the restriction of the bundle is trivial there is a trivialising map $h : U \times G \to \pi^{-1}(U)$ that is equivariant with respect to the $G$ action on $U \times G$ and $\pi^{-1}(U)$. All the specific principal bundles considered above are locally trivial, although this is not always a minor matter to prove: for example, it is a famous theorem[11] that $H \to G \to G/H$ is locally trivial when $G$ is a finite-dimensional Lie group.

2. It follows from the above that the set of all principal maps from a principal $G$-bundle $\xi = (P, \pi, \mathcal{M})$ to itself forms a group. This is called the *automorphism group* of the bundle and is denoted $\mathrm{Aut}(\xi)$. In Yang-Mills theory in a non-trivial instanton sector, the automorphism group of the bundle plays the same role as the familiar group of 'gauge transformations' in the instanton-zero sector (when

[10] See Husemoller (1966) for full proofs of results of this type, including showing that the maps involved are continuous.

[11] For example, see Chevalley (1946).

the bundle is trivial). Indeed, as shown by the following theorem, when $\xi$ is trivial, $\mathrm{Aut}(\xi)$ is isomorphic to the group $C^\infty(\mathcal{M}, G)$ of smooth gauge transformations.

**Theorem 5.2** *Let $\xi$ be the product bundle $(\mathcal{M} \times G, \mathrm{pr}_1, \mathcal{M})$. Then the automorphisms $u : \mathcal{M} \times G \to \mathcal{M} \times G$ of this principal bundle are in one-to-one correspondence with maps $\chi : \mathcal{M} \to G$ such that $u(x, g) = (x, \chi(x)g)$.*

**Proof**

Let $u$ be an automorphism of $\xi$. Then $u(x, g) = (x, f(x, g))$ for some $f : \mathcal{M} \times G \to G$. Define a map $\chi_u : \mathcal{M} \to G$ by $\chi_u(x) := f(x, e)$. Then $u(x, g) = (u(x, e))g = (x, \chi_u(x))g = (x, \chi_u(x)g)$ where I have used the obvious right $G$-action on the product space $\mathcal{M} \times G$.

Conversely, given $\chi$ define $u_\chi(x, g) := (x, \chi(x)g)$. Then we have $(u_\chi(x, g))g' = (x, \chi(x)gg') = u_\chi(x, gg')$. **QED**

It has been remarked already that one of the significant features of fibre bundles is the way in which—like differential forms—they can be pulled back via a map between a pair of manifolds. The next result is important since it shows that the property of being a principal bundle is preserved in this process.

**Theorem 5.3** *Let $\xi = (P, \pi, \mathcal{M})$ be a principal $G$-bundle and let $f : \mathcal{M}' \to \mathcal{M}$. Then the induced bundle $f^*(\xi)$ is also a principal $G$-bundle.*

**Proof**

Let $f^*(\xi) = (P', \pi', \mathcal{M}')$ and consider the diagram

$$\begin{array}{ccccc} & & P' & \xrightarrow{f_\xi} & P \\ & & \downarrow{\scriptstyle \pi'} & & \downarrow{\scriptstyle \pi} \\ P'/G & \xrightarrow{h} & \mathcal{M}' & \xrightarrow{f} & \mathcal{M} \end{array} \tag{5.2.14}$$

Define a $G$-action on $\mathcal{M}' \times P$ by $(x', p)g := (x', pg)$. This induces an action on $P'$ since $\pi(pg) = \pi(p) = f(x')$. Clearly this action is free. Now define $h : P'/G \to \mathcal{M}'$ by $h((x',p)_G) := \pi'((x',p)_G) = x'$ where $(x',p)_G$ denotes the $G$-orbit that passes through $(x',p) \in P'$ with $f(x') = \pi(p)$. Then

(i) The map $h$ is well defined since if $(x',p)_G = (x',q)_G$ then there exists $g \in G$ such that $q = pg$, but then $h(x',q)_G = h(x',pg)_G = \pi'((x',pg)_G) = x' = h(x',p)_G$, as required.

(ii) $h$ is injective, since if $h((x'_1,p_1)_G) = h((x'_2,p_2)_G)$ then $x'_1 = x'_2$, and hence $\pi(p_1) = f(x'_1) = f(x'_2) = \pi(p_2)$ which implies that there exists $g \in G$ such that $p_1 = p_2 g$. which means precisely that $(x'_1,p_1)_G = (x'_2,p_2)_G$.

(iii) $h((x',p)_G) := x'$ clearly defines a map that is surjective.

(iv) $h$ and $h^{-1}$ are continuous [Exercise!].

Thus the pull-back bundle $f^*(\xi) = (P',\pi',\mathcal{M}')$ is a principal $G$-bundle. **QED**

**Comments**

1. The theorem shows in particular that the restriction $\xi_{|U}$ of $\xi$ to a subspace $U$ of $\mathcal{M}$ is itself a principal bundle.

2. Since the twists of any fibre bundle are inherited from the parent principal bundle, it follows that particular interest is attached to the universal $G$-bundles $G \to EG \to BG$. As for all universal bundles, these are characterised by the property that $EG$ is a *contractible* space.

Thus, for example, the infinite covering of $S^1$ by the real line under the map $\pi : \mathbb{R} \to S^1$, $r \mapsto e^{ir}$, produces a bundle $\mathbf{Z} \to \mathbb{R} \to S^1$ that is a universal $\mathbf{Z}$-bundle. Thus the principal $\mathbf{Z}$-bundles over any manifold $\mathcal{M}$ are classified by the homotopy set $[\mathcal{M}, S^1]$. Now $\pi_1(S^1) \simeq \mathbf{Z}$ and all other homotopy groups of $S^1$ vanish; *i.e.*, $S^1$ is an Eilenberg-Maclane[12] space $K(\mathbf{Z},1)$. Hence $[\mathcal{M}, S^1] \simeq H^1(\mathcal{M}, \mathbf{Z})$, giving rise to

[12]In general, if $\pi$ is any group, and if $n$ is any positive integer, an *Eilenberg-Maclane space* $K(\pi,n)$ is defined to be any topological space with the property that $\pi_n(K(\pi,n)) \simeq \pi$ and with all other homotopy groups vanishing. It can

the classification result

$$\mathcal{B}_{\mathbf{Z}}(\mathcal{M}) \simeq H^1(\mathcal{M}, \mathbf{Z}). \tag{5.2.15}$$

3. A more sophisticated example is afforded by the sequence of principal $U1$-bundles (see Eq. (5.2.7)) $U1 \to S^{2n+1} \to \mathbb{C}P^n$, $n = 1, 2, \ldots$, in which the $U1$ group is regarded as the group of all complex numbers of modulus one, and where $e^{i\theta} \in U1$ acts on an element $(z_1, z_2, \ldots, z_n)$ in $S^{2n+1}$ (identified as the set of all vectors $\vec{z} \in \mathbb{C}^{n+1}$ such that $\sum_{i=1}^{n+1} |z_i|^2 = 1$) as

$$(z_1, z_2, \ldots, z_n)e^{i\theta} := (e^{i\theta} z_1, e^{i\theta} z_2, \ldots, e^{i\theta} z_n). \tag{5.2.16}$$

Taking the limit $n \to \infty$ in an appropriate sense, we get the bundle

$$U1 \to S^\infty \to \mathbb{C}P^\infty. \tag{5.2.17}$$

Now, the fact that $\pi_i(S^n) \simeq 0$ for $i < n$, generalises to the feature that all the homotopy groups of $S^\infty$ vanish, which—since it is a CW complex—implies that $S^\infty$ is a contractible space. Thus Eq. (5.2.17) is a universal $U1$-bundle, and hence $U1$-bundles over an arbitrary manifold $\mathcal{M}$ are classified by the homotopy set $[\mathcal{M}, \mathbb{C}P^\infty]$. A similar argument to that employed for the infinite-sphere, shows that $\pi_2(\mathbb{C}P^\infty) \simeq \mathbf{Z}$ and that all other homotopy groups vanish; *i.e.*, $\mathbb{C}P^\infty$ is an Eilenberg-Maclane space $K(\mathbf{Z}, 2)$. This in turn implies the famous classification result

$$\mathcal{B}_{U1}(\mathcal{M}) \simeq H^2(\mathcal{M}, \mathbf{Z}). \tag{5.2.18}$$

4. The set of Yang-Mills potentials $\mathcal{A}$, and the set of Riemannian metrics Riem$\mathcal{M}$ are both contractible spaces. Thus the fundamental 'gauge-bundles' $\mathcal{G} \to \mathcal{A} \to \mathcal{A}/\mathcal{G}$ and Diff$\mathcal{M} \to$ Riem$\mathcal{M} \to$ Riem$\mathcal{M}$/Diff$\mathcal{M}$ are universal bundles for the gauge group $\mathcal{G}$ and the diffeomorphism group Diff$\mathcal{M}$ respectively.

---

be shown that (i) any two Eilenberg-Maclane spaces of type $(\pi, n)$ are homotopically equivalent; and (ii) the homotopy set $[\mathcal{M}, K(\pi, n)]$ is equal to the singular cohomology group $H^n(\mathcal{M}, \pi)$ of $\mathcal{M}$ with coefficients in the group $\pi$.

### 5.2.3 Cross-sections of a principal bundle

We come now to a theorem that is of considerable significance in a variety of applications of fibre bundle theory in differential geometry and theoretical physics, especially in those areas where global topological properties are important. This theorem states that—contrary to what is sometimes felt to be intuitively 'obvious'—a principal fibre bundle does not have any smooth cross-sections unless it is untwisted! More precisely:

**Theorem 5.4** *A principal $G$-bundle $(P, \pi, \mathcal{M})$ is trivial (*i.e., *$\mathcal{M}$-isomorphic to $(\mathcal{M} \times G, \mathrm{pr}_1, \mathcal{M})$) if and only if it possesses a continuous cross-section.*

**Proof**

Let $\sigma : \mathcal{M} \to P$ be a cross-section of the bundle. Then, for any point $p \in P$, there must exist $\chi_\sigma(p) \in G$ such that $p = \sigma(\pi(p))\chi_\sigma(p)$. The resulting function $\chi_\sigma : P \to G$ is unique because the $G$-action on the bundle space $P$ is free. Clearly $\chi_\sigma(pg) = \chi_\sigma(p)g$. Now define $u_\sigma : P \to \mathcal{M} \times G$ by $u_\sigma(p) := (\pi(p), \chi_\sigma(p))$. Then,

(i) $u_\sigma$ is a bundle map; and

(ii) $u_\sigma(pg) = (\pi(p), \chi_\sigma(pg)) = (\pi(p), \chi_\sigma(p)g) = (\pi(p), \chi_\sigma(p))g = u_\sigma(p)g$.

Hence $u_\sigma$ is a principal map and therefore, by the theorem above, an isomorphism.

The converse statement is trivial. Namely, if $h : \mathcal{M} \times G \to P$ is a principal map then we can define a cross-section $\sigma_h : \mathcal{M} \to P$ by $\sigma_h(x) := h(x, e)$.

The proof of the theorem is completed by showing that all the various maps are continuous (or smooth, as appropriate); this is left as an exercise. **QED**

**Comments**

1. The inverse of the map $u_\sigma : P \to \mathcal{M} \times G$ is $h_\sigma : \mathcal{M} \times G \to P$ defined by $h_\sigma(x, g) := \sigma(x)g$.

2. It is clearly of some importance to know when a principal bundle is trivial. This is primarily a global problem and involves a delicate interplay between the topological structures of the fibre and base space. One well-known result is that every principal bundle defined over a contractible, paracompact[13] base space is necessarily trivial (see later).

3. An example of a non-trivial principal bundle arises in Yang-Mills theory and is responsible for the so-called 'Gribov effect'. This is the bundle $\mathcal{G} \to \mathcal{A} \to \mathcal{A}/\mathcal{G}$ referred to earlier in which $\mathcal{A}$ is the set of all Yang-Mills potentials and $\mathcal{G}$ is the gauge group of the theory. Choosing a cross-section of this bundle corresponds physically to choosing a *gauge*. However, as Singer showed, this particular principal bundle is definitely not trivial and hence no smooth cross-sections exist (Singer 1978). A similar situation arises in the corresponding situation in general relativity with the principal bundle $\text{Diff}\mathcal{M} \to \text{Riem}\mathcal{M} \to \text{Riem}\mathcal{M}/\text{Diff}\mathcal{M}$. The Gribov-Singer effect means that there is an intrinsic obstruction to choosing a gauge that works for all physical configurations; whether or not this has any real significance in the theory is still a matter for debate.

4. The precise extent to which a global cross-section of a principal bundle does *not* exist can be used to give a topological classification of the twists in the bundle, and hence to relate to the theory of characteristic classes. One approach (which is applicable for any fibre $F$) involves triangulating the base space sufficiently finely that the bundle is trivial over any simplex, and then trying to construct a cross-section by first defining it on the 0-simplices (this is always possible since the 0-simplices are just points) then extending it to the 1-simplices, then to the 2-simplices, and so on. At some stage,

[13] A Hausdorff space $X$ is *paracompact* if for every open covering of $X$ there exists another covering that refines the first. The importance in differential geometry of paracompact spaces lies in the existence of a *partition of unity* for every locally-finite covering of $X$. Roughly speaking, a partition of unity is a collection of positive, real-valued smooth functions, one for each open set of the covering and which vanishes outside the open set; and with the property that the sum of the functions is 1 everywhere. Partitions of unity are used to patch together locally defined geometrical objects to give a global object. For details see, for example, (Helgason 1962).

if the bundle is non-trivial, a dimension $n$ will be reached for which the cross-section cannot be extended to all the simplices of the next higher dimension. Thus there will be some $n$-simplex $\Delta$ and a map $s$ from $\Delta$ into the fibre $F$ which cannot be extended to the interior of $\Delta$. But $\Delta$ is homeomorphic to an $n$-sphere, and the inability to extend $s$ is equivalent to the statement that the element in $\pi_n(F)$ corresponding to the homotopy class of $s$ is non-trivial.

It can be shown that this association to $\Delta$ of an element of $\pi_n(F)$ depends only on the cohomology class of the $(n+1)$-cell interior of $\Delta$, and hence generates an element of the cohomology group $H^{n+1}(\mathcal{M}, \pi_n(F))$. This gives the *primary obstruction* to constructing a continuous section of the bundle. In general, this analysis shows that potential obstructions to constructing a cross-section lie in the cohomology groups $H^k(\mathcal{M}, \pi_{k-1}(F))$, $k = 1, 2, \ldots$. In particular, this analysis can be applied to a universal principal bundle $G \to EG \to BG$ and yields the basic universal characteristic classes for all principal $G$-bundles. For discussion of this and related topics see Steenrod (1951).

# 5.3 Associated Bundles

## 5.3.1 The main definition

In this section we shall discuss a method of constructing a great variety of fibre bundles that are associated in a precise way with some principal bundle. The basic idea is that given a particular principal bundle $(P, \pi, \mathcal{M})$ with structure group $G$, we can form a fibre bundle with fibre $F$ for each space $F$ on which $G$ acts as a group of transformations. First we must introduce the concept of a '$G$-product' of a pair of spaces on which $G$ acts.

**Definition 5.11**

1. Let $X$ and $Y$ be any pair of right $G$-spaces. Then the *$G$-product* of $X$ and $Y$ is the space of orbits of the $G$-action on the Cartesian product $X \times Y$. Thus we define an equivalence relation on $X \times Y$ in which $(x, y) \equiv (x', y')$ if there exists $g \in G$ such that $x' = xg$ and $y' = yg$.

2. The $G$-product is denoted $X \times_G Y$, and the equivalence class of $(x, y) \in X \times Y$ is written as $(x, y)_G$, or just $[x, y]$ if the group $G$ is clear.

Note that, given any right $G$-space $Y$, the group $G$ can itself be used as a right $G$-space $X$ to give the $G$-product $G \times_G Y$. It is relevant for our considerations later that this space is *homeomorphic* to $Y$:

**Lemma** The map $\iota : G \times_G Y \to Y$ defined by $\iota([g, y]) := yg^{-1}$ is a homeomorphism.

**Proof**

(i) This map is well-defined since $[g_1, y_1] = [g_2, y_2]$ implies that there exists $g \in G$ such that $g_2 = g_1 g$ and $y_2 = y_1 g$. Therefore $y_2 g_2^{-1} = y_1 g g^{-1} g_1^{-1} = y_1 g_1^{-1}$.

(ii) The map $\iota$ is surjective because $\iota([e, y]) = y$.

(iii) The map $\iota$ is injective since if $\iota([g_1, y_1]) = \iota([g_2, y_2])$ then $y_1 g_1^{-1} = y_2 g_2^{-1}$ and so $(g_2, y_2) = (g_1 g_1^{-1} g_2, y_1 g_1^{-1} g_2)$, and hence $(g_2, y_2) \equiv (g_1, y_1)$.

Finally, it should be shown that the map $\iota$ and its inverse are continuous. As usual, this is left as an exercise (or see Husemoller (1966)), as is the corresponding proof of differentiability in the manifold case with $G$ being a Lie group. **QED**

Now we come to the crucial definition of an 'associated fibre bundle'.

**Definition 5.12**

Let $\xi = (P, p, \mathcal{M})$ be a principal $G$-bundle and let $F$ be a left $G$-space. Define $P_F := P \times_G F$ where $(p, v)g := (pg, g^{-1}v)$, and define a map $\pi_F : P_F \to \mathcal{M}$ by $\pi_F([p, v]) := \pi(p)$. Then $\xi[F] := (P_F, \pi_F, \mathcal{M})$ is a fibre bundle over $\mathcal{M}$ with fibre $F$ that is said to be *associated* with the principal bundle $\xi$ via the action of the group $G$ on $F$.

Note that $\pi_F$ is well-defined since if $[p_1, v_1] = [p_2, v_2]$ then there exists $g \in G$ with $(p_2, v_2) = (p_1 g, g^{-1} v_1)$ and then $\pi(p_2) = \pi(p_1 g) = \pi(p_1)$.

Note also that in order for the definition to make sense, we must show that $\xi[F]$ really *is* a fibre bundle. The crucial step is contained in the next theorem.

**Theorem 5.5** *For each $x \in \mathcal{M}$, the space $\pi_F^{-1}(\{x\})$ is homeomorphic to $F$.*

**Proof**

To each point $p \in P(\xi)$ there is associated the map $\iota_p : F \to P_F$ defined by $\iota_p(v) := [p, v]$. Now, $\pi_F[p, v] = \pi(p)$ and hence $\iota_p(F) \subset \pi_F^{-1}(\{\pi(p)\})$. Choose some particular point $p_0 \in \pi^{-1}(\{x\})$ so that $\iota_{p_0}(F) \subset \pi_F^{-1}(\{x\})$. Now, if $[p, v] \in \pi_F^{-1}(\{x\})$ define $j_{p_0} : \pi_F^{-1}(\{x\}) \to F$ as the map that takes $[p, v]$ into $\tau(p_0, p)v$ where $\tau(p_0, p)$ is the so-called *translation function* of the principal bundle $\xi$, whose value is defined as the unique element in $G$ such that $p = p_0\tau(p_0, p)$. The map $j_{p_0}$ is well-defined since $\tau(p_0, pg)g^{-1}v = \tau(p_0, p)v$. But,

(i) $\iota_{p_0} \circ j_{p_0}([p, v]) = [p_0, \tau(p_0, p)v] = [p_0\tau(p_0, p), v] = [p, v]$

(ii) $j_{p_0} \circ \iota_{p_0}(v) = j_{p_0}([p_0, v]) = \tau(p_0, p_0)v = v.$

Hence the maps $\iota_{p_0}$ and $j_{p_0}$ are inverses of each other, and therefore $j_{p_0}$ is a homeomorphism (see Husemoller (1966) for a discussion of the continuity properties of the translation function). **QED**

**Comments**

1. It is clear that the 'twists' in an associated bundle $\xi[F]$ are determined by the twists in the underlying principal bundle $\xi = (P, \pi, \mathcal{M})$ and also by the way in which the group $G$ acts on the fibre $F$. In particular, if $G$ acts trivially on $F$ (*i.e.*, $gv = v$ for all $g \in G$ and all $v \in F$) then, considered only as a *bundle*, $\xi[F]$ is $\mathcal{M}$-isomorphic to the product bundle $(\mathcal{M} \times F, \mathrm{pr}_1, \mathcal{M})$ via the map $[p, v] \mapsto (\pi(p), v)$.

However, within the category of *associated* bundles, $\xi[F]$ is only deemed to be trivial if the underlying bundle $\xi$ is trivial as a principal $G$-bundle; see below for further remarks on this potentially confusing nomenclature. □

**Examples**

1. Let $\xi$ be the principal $\mathbf{Z}_2$-bundle formed from the action of the group $\mathbf{Z}_2 = \{e, a\}$ on the circle $S^1$, in which $a \in \mathbf{Z}_2$ sends $x \in S^1$ into $-x$. The base space $S^1/\mathbf{Z}_2$ is also diffeomorphic to $S^1$, and $\xi$ is simply the double covering of the circle by a circle that was discussed earlier in the context of Figure 5.7.

A number of interesting bundles can formed from this particular principal bundle by the process of association. For example:

(a) Let $\mathbf{Z}_2$ act on $F = [-1, 1]$ by $a \in \mathbf{Z}_2$ taking $r \in [-1, 1]$ into itself (*i.e.*, the trivial action). Then the associated bundle $\xi[F]$ is just a cylinder; *i.e.*, considered solely as a bundle it is isomorphic to the product space $S^1 \times [-1, 1]$.

(b) Let $\mathbf{Z}_2$ act non-trivially on $[-1, 1]$ by $a \in \mathbf{Z}_2$ taking $r \in [-1, 1]$ into $-r$. Then the associated bundle $\xi[F]$ is the Möbius strip.

(c) Let $\mathbf{Z}_2$ act on $F = S^1$ by $a \in \mathbf{Z}_2$ reflecting the elements of the circle in a diameter. In this case the ensuing associated bundle is the Klein bottle.

Thus we see that some of the most familiar examples of fibre bundles are obtained from what is arguably the simplest non-trivial principal bundle.

2. The next example is of major importance in differential geometry as it provides a 'bundle-theoretic' way of viewing general tensorial structures. In this case, the principal bundle of interest is the bundle of frames $\mathbf{B}(\mathcal{M})$ on the $m$-dimensional manifold $\mathcal{M}$. The structure group is $GL(m, \mathbb{R})$, which can act on a variety of spaces, especially vector spaces.

A simple case is when $F = \mathbb{R}^m$ and the action of $GL(m, \mathbb{R})$ is just the usual linear group of transformations. Then the associated bundle $\mathbf{B}(\mathbb{R}^m) = \mathbf{B}(\mathcal{M}) \times_{GL(m,\mathbb{R})} \mathbb{R}^m$ can be identified with the tangent bundle $T\mathcal{M}$ via the map

$$[b, \vec{r}] \mapsto \sum_{i=1}^{m} b_i r^i \in T_x\mathcal{M} \qquad (5.3.1)$$

where $b$ is a base for the tangent space $T_x\mathcal{M}$ (*i.e.*, $b \in \mathbf{B}_x(\mathcal{M})$) and $\vec{r} = (r^1, r^2, \ldots, r^m)$.

3. More generally, let $\rho : GL(m, \mathbb{R}) \to \mathrm{Aut}V$ be any representation of $GL(m, \mathbb{R})$ as a group of linear transformations of a real vector space $V$. Then $\mathbf{B}[V]$ is called the *bundle of tensors of type* $\rho$ (it is actually a vector bundle—see below), and cross-sections of this vector bundle are *tensor fields of type* $\rho$. For example:

(a) Let $V = \mathbb{R}$ and let $\rho$ be the identity transformation $\rho(a) = 1$ for all $a \in GL(m, \mathbb{R})$. Then the corresponding associated bundle is the bundle of absolute scalars. The bundle space is diffeomorphic to the product $\mathcal{M} \times \mathbb{R}$ and the cross-sections are in one-to-one correspondence with the usual real-valued functions on $\mathcal{M}$.

(b) Now let $V = \mathbb{R}$ again but this time set $\rho(a) = (\det a)^\omega$ for some $\omega \in \mathbb{R}$. The ensuing associated bundle is called the bundle of *scalar densities of weight* $\rho$.

(c) More generally let $V = \mathbb{R}^m \otimes \mathbb{R}^m \otimes \cdots \otimes \mathbb{R}^m \otimes (\mathbb{R}^m)^* \otimes (\mathbb{R}^m)^* \otimes \cdots \otimes (\mathbb{R}^m)^*$ where the first tensor product is taken $k$ times and the second $l$ times, and let $a \in GL(m, \mathbb{R})$ act on $\vec{v} \in V$ by

$$(\rho(a)\vec{v})^{i_1\ldots i_l}_{j_1\ldots j_k} := (\det a)^\omega \sum_{k_1\ldots k_l=1}^{m} \sum_{h_1\ldots h_k=1}^{m} a_{i_1k_1} \ldots a_{i_lk_l} a_{h_1j_1} \ldots a_{h_kj_k} v^{k_1\ldots k_l}_{h_1\ldots h_k} \tag{5.3.2}$$

The associated bundle is known as the bundle of *tensor densities of weight* $\omega$*, contravariant of rank* $k$ *and covariant of rank* $l$. This example is quite useful in so far as the definition of densities in a 'non-bundle' way tends to be a little awkward.

### 5.3.2 Associated bundle maps

At various places above, we have implied that the total space $\xi[F]$ of an associated bundle is 'like' some other space $Z$ that is well known. Generally speaking, we mean by this merely that $\xi[F]$ is diffeomorphic to $Z$, or is isomorphic to $Z$ as a bundle. But this raises the question of is meant by a genuine 'morphism' between two associated bundles. Of course, this could be defined simply as a bundle map, but this would not take into account the 'associated' status of the bundles and the corresponding relations to their underlying principal bundles. The most useful definition is to start with a principal bundle map between these underlying bundles, and then use this to construct a bundle map between the two associated bundles. Not every bundle map is necessarily of this type, and it follows therefore that this definition is more restrictive than the first one.

The critical definitions are as follows:

**Definition 5.13**

1. Let $(u, f)$ be a principal bundle map between a pair of principal $G$-bundles $\xi = (P, \pi, \mathcal{M})$ and $\xi' = (P', \pi', \mathcal{M}')$. A map between the associated bundles $P \times_G F$ and $P' \times_G F$ can be defined by $u_F([p, v]) := [u(p), v]$. This is well-defined since

   $$u_F([pg, g^{-1}v]) = [u(pg), g^{-1}v] = [u(p)g, g^{-1}v] = [u(p), v], \tag{5.3.3}$$

   and it is clearly a bundle map [Exercise!].

   An *associated bundle map* between a pair of associated bundles $\xi[F]$ and $\xi'[F]$ is defined to be a bundle map of the type above.

2. An associated fibre bundle $\xi[F]$ is *trivial* if its underlying principal bundle $\xi$ is trivial. Note that this implies that $\xi[F]$ is trivial as a bundle, but the converse is false. For example (as remarked above), if $\xi$ is any non-trivial principal $G$-bundle and $F$ is any space on which $G$ acts trivially, then—as a bundle—$\xi[F]$ is trivial but it is not trivial as an associated bundle. A failure to observe this subtle distinction can cause some confusion when reading the literature.

3. Two associated bundles $\xi$[F] and $\zeta[F]$ are said to be *locally isomorphic* if the underlying principal bundles are locally isomorphic, *i.e.*, if for each $x \in \mathcal{M}$, there exists an open neighbourhood $U \subset \mathcal{M}$ of $x$ such that $\xi_{|U}$ and $\zeta_{|U}$ are isomorphic as principal bundles.

We consider now the important question of the status of the pull-back of an associated fibre bundle. We know that it is a bundle, but is it associated to a principal bundle? The answer is contained in the following theorem:

**Theorem 5.6** *Let $\xi = (P, \pi, \mathcal{M})$ be a principal $G$-bundle with base space $\mathcal{M}$ and let $G$ act on a space $F$ to produce the associated bundle $\xi[F]$. Let $\mathcal{M}'$ be some other space and let $f : \mathcal{M}' \to \mathcal{M}$. Then there is a bundle $\mathcal{M}'$-isomorphism between (i) the pull-back $f^*(\xi[F])$ of $\xi[F]$, and (ii) the bundle obtained by first pulling back $\xi$ to give the principal bundle $f^*(\xi)$ and then forming the associated bundle $f^*(\xi)[F]$.*

**Proof**

Construct a map $j : f^*(\xi[F]) \subset \mathcal{M}' \times (P \times_G F) \to f^*(\xi)[F] \subset (\mathcal{M}' \times P) \times_G F$ by $j(x, [p, v]) := [(x, p), v]$. Then it is a straightforward exercise [!] to show that $j$ is an isomorphism. **QED**

A simple corollary to the above is the result that the restriction $\xi[F]_{|\mathcal{N}}$ of the associated bundle $\xi[F]$ to a subspace $\mathcal{N} \subset \mathcal{M}$ is $\mathcal{N}$-isomorphic (as a bundle) to the bundle $\xi_{|\mathcal{N}}[F]$ which is associated to the restriction $\xi_{|\mathcal{N}}$ to $\mathcal{N}$ of the principal bundle $\xi$.

**Comments**

1. Let $\xi = (\mathcal{M} \times G, \mathrm{pr}_1, \mathcal{M})$ and let $F$ be any $G$-space. Then the associated bundle $\xi[F]$ is $\mathcal{M}$-isomorphic to the product bundle $(\mathcal{M} \times F, \mathrm{pr}_1, \mathcal{M})$ with a map $\iota : (\mathcal{M} \times G)_G \times F \to \mathcal{M} \times F$ defined by $\iota([(x, g), v]) := (x, gv)$.

2. As a bundle, $\xi[F]$ is locally trivial. To see this note that a principal bundle $\xi$ is locally trivial, and hence, for each $x \in \mathcal{M}$, there exists an open neighbourhood $U \subset \mathcal{M}$ of $x$ and a bundle isomorphism $h : U \times G \to P(\xi)_{|U}$. If $F$ is a $G$-space, we can define $h' : U \times F \to P_{F|U}$ by

$$h'(x, v) := [h(x, e), v] \tag{5.3.4}$$

which is clearly a bundle $U$-map. (N.B. If $h$ comes from a local section $\sigma : U \to P(\xi)$ then $h'(x, v) = [\sigma(x), v]$.) Similarly, we can define a map $u' : P_{F|U} \to U \times F$ by

$$u'([h(x, g), v]) := (x, gv). \tag{5.3.5}$$

It is easy to see that $h' \circ u' = \mathrm{id}_{P_{F|U}}$ and $u' \circ h' = \mathrm{id}_{U \times F}$. Thus $h'$ defines a bundle $U$-isomorphism from $U \times F$ to $P_{F|U}$; hence, as a bundle, $\xi[F]$ is locally trivial.

3. If $U_1$ and $U_2$ are a pair of neighbourhoods of $x \in \mathcal{M}$ over which $\xi$ is trivial, there will exist a pair of corresponding maps $h'_{U_1} : U_1 \times F \to P_{F|U_1}$ and $h'_{U_2} : U_2 \times F \to P_{F|U_2}$ together with the pair of principal bundle local trivialisations $h_{U_1} : U_1 \times G \to P_{|U_1}$ and $h_{U_2} : U_2 \times G \to P_{|U_2}$. On the overlap $U_1 \cap U_2$ there is therefore a principal bundle morphism $h_{U_2}^{-1} \circ h_{U_1} : (U_1 \cap U_2) \times G \to (U_1 \cap U_2) \times G$. Thus there exists a map $g_{U_1U_2} : U_1 \cap U_2 \to G$ such that $h_{U_2}^{-1} \circ h_{U_1}(x, g) = (x, g_{U_2U_1}(x)g)$. These functions $g_{U_iU_j}$ are known as the *structure functions*, or *transition functions* of the bundle. They satisfy the relations

$$\text{(a)} \quad g_{U_1U_1}(x) = e \tag{5.3.6}$$

$$\text{(b)} \quad g_{U_1U_2}(x) = (g_{U_2U_1}(x))^{-1} \tag{5.3.7}$$

$$\text{(c)} \quad g_{U_1U_2}(x)\, g_{U_2U_3}(x) = g_{U_1U_3}(x) \tag{5.3.8}$$

for all $U_1$, $U_2$, $U_3$ with $U_1 \cap U_2 \cap U_3 \neq \emptyset$. These relations are characteristic in the sense that any family of $G$-valued functions satisfying these relations are necessarily the transition functions of some fibre bundle. This is the foundation of the approach to fibre bundle theory based on local bundle coordinate charts—see Steenrod (1951) for further details.

4. According to the definition above, the natural definition of an *automorphism* of an associated bundle $\xi[F]$ is a bundle map $u_F$ defined by $u_F([p, v]) := [u(p), v]$ where $u \in \mathrm{Aut}\,\xi$ is an automorphism of the principal bundle $\xi$. In particular, if $\xi = (\mathcal{M} \times G, \mathrm{pr}_1, \mathcal{M})$ is the product bundle, every automorphism of $\xi$ is of the form $u(x, g) = (x, \chi(x)g)$ for some map $\chi : \mathcal{M} \to G$. Then, on the associated bundle $\xi[F]$ we have $u_F([(x, g), v]) = [(x, \chi(x)g), v]$ or, using the bundle-isomorphism of $P_F$ with $\mathcal{M} \times F$ defined above—$\iota([(x, g), v]) = (x, gv)$—we can write $u_F$ as the map from $\mathcal{M} \times F$ into itself defined by $(x, v) \mapsto (x, \chi(x)v)$. This is relevant to understanding how a gauge transformation in Yang-Mills theory acts on the matter fields in the system.

5. Let $\xi$ be any principal $G$-bundle. Then $G$ is itself a left $G$-space under left multiplication. The ensuing associated bundle $\xi[G]$ is $\mathcal{M}$-isomorphic to $\xi$ by the map defined by $i : P \times_G G \to P$ with $i([p, g]) := pg$.

More generally, if $H$ is a subgroup of $G$ and if $\eta$ is a principal $H$-bundle, we can form the associated bundle $\eta[G] = P \times_H G$ which is a bundle with fibre $G$. As we shall see below, it is in fact a *principal* $G$-bundle.

6. If $\xi = (P, \pi, \mathcal{M})$ is any principal $G$-bundle, a very important class of associated bundles is obtained by considering the action of $G$ on the coset space $G/H$, where $H$ is a closed subgroup of $G$, and hence constructing $P \times_G G/H$. This is closely related to the bundle $\xi/H$ whose total space is defined to be the orbit space $P/H$ of the $H$-action on $P(\xi)$ and whose projection map onto $\mathcal{M}$ is defined as $\pi'([p]_H) := \pi(p)$. There is an $\mathcal{M}$-isomorphism $j : P \times_G G/H \to P/H$ defined by

$$j((p, [g]_H)_G) := [pg]_H \tag{5.3.9}$$

whose inverse $k : P/H \to P \times_G H$ is given by

$$k([p]_H) := (p, [e]_H)_G. \tag{5.3.10}$$

Note that, in physical terms, the cross-sections of such a bundle are the fields for a 'twisted' non-linear $\sigma$-model.

### 5.3.3 Restricting and extending the structure group

The example above—where a principal $H$-bundle $\eta$ was converted into a bundle $\eta[G]$ with fibre $G$—is of considerable importance and merits further discussion.

**Definition 5.14**

Let $H$ be a closed subgroup of $G$ and let $\eta$ and $\xi$ be a principal $H$-bundle and $G$-bundle respectively, defined over the same base space. Suppose there exists a bundle map $u$ from $P(\eta)$ to $P(\xi)$ that is equivariant with respect to the subgroup $H$, *i.e.*, $u(ph) = u(p)h$ for all $h \in H \subset G$. Then,

(a) $\eta$ is called a *H-restriction* of $\xi$;

(b) $\xi$ is called a *G-extension* of $\eta$.

**Comments**

1. Let $\eta = (P, \pi, \mathcal{M})$ be a principal $H$-bundle. Then there always exists a $G$-extension of $\eta$ for any group $G$ that contains $H$ as a closed subgroup: simply form the bundle $P \times_H G$ mentioned above and note that this can be given a $G$-bundle structure by defining $[p, g_0]g := [p, g_0 g]$. Then a principal map is defined by $u : P \to P \times_H G$, $u(p) := [p, e]$.

2. Let $\xi$ be any principal $G$-bundle with an $H$-restriction $\eta$. Then there is a principal isomorphism of $\xi$ with $\eta[G]$ defined by $\iota : P(\xi) \to P(\eta) \times_H G$, $\iota(p) := [q, g]$ where $(q, g) \in P(\eta) \times G$ is any pair of elements such that $p = u(q)g$ (the map $\iota$ does not depend on the choice) and where $u : P(\eta) \to P(\xi)$ is the embedding map of $\eta$ in $\xi$. Thus the $G$-extension of an $H$-restriction of a bundle is isomorphic to the bundle.

3. More generally, if $\xi$ is a principal $G$-bundle with an $H$-restriction $\eta$, and if $F$ is a left $G$-space (and hence, *a fortiori*, a left $H$-space), there is a bundle $\mathcal{M}$-map of $\eta[F]$ with $\xi[F]$ in which $(q, v)_H$ is mapped to $(q, v)_G$. □

Now we must return to the non-trivial matter of deciding when a principal $G$-bundle admits a restriction to any specific subgroup $H \subset G$. This question is of considerable physical interest. For example, in Yang-Mills theory it is related to the question of the spontaneous breakdown of the internal symmetry group from $G$ to $H$. In differential geometry, reductions of the structure group $GL(m, \mathbb{R})$ of the frame bundle $\mathbf{B}(\mathcal{M})$ to the subgroup $O(m, \mathbb{R})$ are in one-to-one correspondence with Riemannian metrics on $\mathcal{M}$ (see below).

A theorem of major importance is the following:

**Theorem 5.7** *A principal G-bundle $\xi = (P, \pi, \mathcal{M})$ has a restriction to a H-bundle if and only if the bundle $\xi/H$ over $\mathcal{M}$ has a cross-section.*

**Proof**

Let $u : P(\eta) \to P(\xi)$ be a restriction map. Then define a cross-section of $\xi/H$ by $\sigma : \mathcal{M} \to P(\xi)/H$, $\sigma(x) := [u(p)]_H$ where $p$ is any element in $\pi_\eta^{-1}(\{x\})$.

Conversely, let $\sigma : \mathcal{M} \to P(\xi)/H \simeq P(\xi) \times_H G/H$ be a cross-section of $\xi/H$:

$$\begin{array}{ccc} & & P(\xi) \\ & & \downarrow{\scriptstyle \rho} \\ \mathcal{M} & \xrightarrow{\sigma} & P(\xi)/H \end{array} \tag{5.3.11}$$

If $\rho(p) := [p]_H$ we have a principal $H$-bundle $\zeta := (P(\xi), \rho, P(\xi)/H)$, and hence $\sigma^*(\zeta)$ is a principal $H$-bundle over $\mathcal{M}$. Thus, as the bundle space of the $H$-bundle $\eta$ being sought, we define $P(\eta) := \{(x,p) \in \mathcal{M} \times P(\xi) \mid \sigma(x) = [p]_H\}$. Finally, define the bundle embedding map $u : P(\eta) \to P(\xi)$ by $u(x,p) := p$. **QED**

**Comments**

1. Assuming that $\mathcal{M}$ is paracompact it can be shown that any fibre bundle whose fibre is a cell (*i.e.*, it is homeomorphic to $\mathbb{R}^n$ for some $n$) always admits a cross-section (for example, see Husemoller (1966)). This follows from the theory discussed briefly in Section 5.2.3 where it was argued that the obstructions to constructing a cross-section of a fibre bundle lie in the cohomology groups $H^k(\mathcal{M}, \pi_{k-1}(F))$, $k = 1, 2, ....$ If $F$ is a cell then all its homotopy groups vanish, and hence so do the potential obstructions.

In particular, if $H$ is the maximal compact subgroup of any non-compact Lie group $G$, then $G/H$ is always a cell. Hence a bundle whose structure group $G$ is non-compact can always be reduced to its maximal compact subgroup. We shall see an important example of this shortly.

2. This example can be generalised in a very significant way. Specifically, it follows from obstruction theory that if $\mathcal{M}$ is an $m$-dimensional differentiable manifold, and if the homotopy groups of $G/H$ satisfy $\pi_i(G/H) \simeq 0$ for all $i = 1, 2, \ldots, m-1$, then the structure group of any principal $G$ bundle over $\mathcal{M}$ can be reduced from $G$ to $H$. If these homotopy groups are non-zero then the potential obstructions

to constructing a cross-section of $\xi[G/H]$—and hence of reducing the group from $G$ to $H$—are represented by elements of the cohomology groups $H^i(\mathcal{M}, \pi_{i-1}(G/H))$, $i = 1, 2, \ldots, m$. (Recall that if $\mathcal{M}$ has dimension $m$ then $H^i(\mathcal{M}) \simeq 0$ for all $i > m$.)

3. An example where the reduction *can* take place arises in considering an 'instanton' Yang-Mills principal bundle $(P, \pi, \mathcal{M})$ over an arbitrary four-dimensional, compact (and therefore 'Euclideanised') spacetime manifold $\mathcal{M}$. Since the cohomology groups $H^i(\mathcal{M})$ vanish when $i > 4$, the only non-vanishing obstructions to reduction must come from $\pi_i(G/H)$, $i = 1, 2, 3$. For example, suppose that $G = SU(3)$ and we are interested in the reduction to the usual $SU(2)$ subgroup. Then we know from Eq. (4.4.33) that $SU(3)/SU(2) \simeq S^5$. But, in general, $\pi_i(S^n) \simeq 0$ for $i < n$, and hence there is no obstruction to reducing the group from $SU(3)$ to $SU(2)$. This is basically the reason why, in four-dimensional spacetimes, there are no topological properties of $SU(3)$ Yang-Mills theories over and above the ones that already arise for the group $SU(2)$.

In this context, notice that—since $SU(2) \simeq S^3$ and $\pi_3(S^3) \simeq \mathbf{Z}$—there *is* a potential obstruction to reducing the structure group from $SU(2)$ to the trivial group $\{e\}$ (*i.e.*, there is an obstruction to constructing a cross-section of the principal $SU(2)$-bundle). This comes from the non-vanishing cohomology group $H^4(\mathcal{M}; \pi_3(S^3)) = H^4(\mathcal{M}; \mathbf{Z}) \simeq \mathbf{Z}$; in fact, the instanton number is precisely the value of this integer.

4. Using the same line of argument one can see that, on a three-dimensional manifold $\Sigma$, there are no obstructions to reducing $SU(2)$ to $\{e\}$. This is of relevance when considering the canonical quantisation of Yang-Mills theory on the physical three-space $\Sigma$. □

### 5.3.4 Riemannian metrics as reductions of $\mathbf{B}(\mathcal{M})$

Now we come to an example of great significance in differential geometry. This shows how Riemannian metrics arise within the context of the bundle of frames.

**Theorem 5.8** *There is a one-to-one correspondence between the Riemannian metrics on an m-dimensional differentiable manifold $\mathcal{M}$ and the reductions of the structure group of the bundle of frames $\mathbf{B}(\mathcal{M})$ from $GL(m,\mathbb{R})$ to $O(m,\mathbb{R})$.*

**Proof**

A *Riemannian metric* $g$ is defined to be a symmetric, positive definite, 2-covariant tensor field. Thus, if $X$, $Y$, $Z$ are any vector fields on $\mathcal{M}$ we have

(i) $g(rX+sY,Z) = rg(X,Z)+sg(Y,Z)$ for all $r,s\in\mathbb{R}$ (5.3.12)

(ii) $g(X,Y) = g(Y,X)$ (5.3.13)

(iii) $g(X,X)\geq 0$ and $=0$ only if $X=0$. (5.3.14)

Suppose first that $g$ is a Riemannian metric on an $m$-dimensional manifold $\mathcal{M}$, and define $\mathbf{O}(\mathcal{M})$ to be the subset of $\mathbf{B}(\mathcal{M})$ made up of all sets of orthonormal basis vectors at all the points of $\mathcal{M}$. This is clearly an $O(m,\mathbb{R})$-bundle (since this subgroup of $GL(m,\mathbb{R})$ preserves the orthonormality) and the required bundle map $\mathbf{O}(\mathcal{M})\to\mathbf{B}(\mathcal{M})$ is simply the subspace embedding.

Conversely, suppose that the group of the principal bundle $\mathbf{B}(\mathcal{M})$ *is* reducible from $GL(m,\mathbb{R})$ to $O(m,\mathbb{R})$ and let $\mathbf{O}(\mathcal{M})$ denote the reduced bundle. For each frame $b\in\mathbf{B}(\mathcal{M})$ there is an injection $\iota_b:\mathbb{R}^m\to B(\mathcal{M})\times_{GL(m,\mathbb{R})}\mathbb{R}^m$ defined by $\iota_b(\vec{r}) := [b,\vec{r}]$. There is also an isomorphism of this associated bundle with the tangent bundle $T\mathcal{M}$ via the map in Eq. (5.3.1), *i.e.*, $[b,\vec{r}]\mapsto\sum_{i=1}^m r^i b_i$. Combining these two maps we can evidently regard $\iota_b$ as a map from $\mathbb{R}^m$ to $T_x\mathcal{M}$, where $x\in\mathcal{M}$ is the point at which $b\in\mathbf{B}(\mathcal{M})$ is a base; *i.e.*, $\iota_b(\vec{r})=\sum_{i=1}^m r^i b_i$.

Now regard $\mathbf{O}(\mathcal{M})$ as a sub-bundle of $\mathbf{B}(\mathcal{M})$. If $u,v\in T_x\mathcal{M}$ choose a base $b\in\pi^{-1}_{\mathbf{O}(\mathcal{M})}(\{x\})$, and define

$$g_x(u,v) := \langle\iota_b^{-1}(u),\iota_b^{-1}(v)\rangle \tag{5.3.15}$$

where $\langle\ ,\ \rangle$ denotes the usual Cartesian inner product on the real vector space $\mathbb{R}^m$. The object Eq. (5.3.15) thus defined is clearly bilinear and positive definite, and the invariance under $O(m, \mathbb{R})$ transformations of the inner product on $\mathbb{R}^m$ means that Eq. (5.3.15) is independent of the particular choice of the base $b$ for $T_x\mathcal{M}$ [Exercise!]. Thus we have acquired a well-defined Riemannian metric on $\mathcal{M}$. **QED**

**Comments**

1. If $b \in \mathbf{B}(\mathcal{M})$ then $\langle \vec{r}_1, \vec{r}_2 \rangle = g_x(\iota_b \vec{r}_1, \iota_b \vec{r}_2) = \sum_{i,j=1}^m r_1^i r_2^j \, g_x(b_i, b_j)$ for all $\vec{r}_1, \vec{r}_2 \in \mathbb{R}^m$.

If $b$ is a base at $x \in \mathcal{M}$ then $b_i = \sum_{\mu=1}^m b_i^\mu(x)(\partial_\mu)_x$. If $b$ is extended to a local section of $\mathbf{B}(\mathcal{M})$ then the set of local vector fields $\{b_1, b_2, \ldots, b_m\}$ is called an *m-bein* and we have, for all $\vec{r}_1, \vec{r}_2 \in \mathbb{R}^m$,

$$\langle \vec{r}_1, \vec{r}_2 \rangle = \sum_{\mu,\nu=1}^m \sum_{i,j=1}^m g_{\mu\nu}(x) b_i^\mu(x) b_j^\nu(x) r_1^i r_2^j \tag{5.3.16}$$

where the *components* $g_{\mu\nu}(x)$ of $g$ with respect to the local coordinate system are

$$g_{\mu\nu}(x) := g_x((\partial_\mu)_x, (\partial_\nu)_x) = g((\partial_\mu)_x \otimes (\partial_\nu)_x). \tag{5.3.17}$$

Hence the components of the metric thus defined are related to the components of the base fields by

$$\delta_{ij} = \sum_{\mu,\nu=1}^m g_{\mu\nu}(x) b_i^\mu(x) b_j^\nu(x) \tag{5.3.18}$$

where $i, j = 1, 2, \ldots, m$.

2. Since $O(m, \mathbb{R})$ is the maximal compact subgroup of $GL(m, \mathbb{R})$ it follows from the earlier discussion of obstruction theory that it is always possible to construct cross-sections of the associated bundle $\mathbf{B}(\mathcal{M}) \times_{O(m,\mathbb{R})} GL(m, \mathbb{R})/O(m, \mathbb{R})$, and hence to find Riemannian metrics on $\mathcal{M}$.

3. We can also consider the reduction of $GL(m, \mathbb{R})$ to the non-compact group $O(m-1, 1)$, which corresponds to finding a Lorentzian signature metric on $\mathcal{M}$. However, the coset space $GL(m, \mathbb{R})/O(m - 1, 1)$ is *not* a cell, and hence there may be topological obstructions

to the construction of such pseudo-Riemannian metrics. In fact, this coset space is homotopic to the real projective space $\mathbb{R}P^{m-1}$ which has many non-vanishing homotopy groups. This topological property was exploited in the theory of 'metric kinks'—one of the earlier examples of the use of topological methods in theoretical physics. □

### 5.3.5 Cross-sections as functions on the principle bundle

Finally we come to a result concerning cross-sections of associated bundles that has been widely used in induced representation theory—in which vectors in the Hilbert space are cross-sections of certain vector bundles—and in studies of anomalies in quantised gauge theories.

**Theorem 5.9** *If $(P_F, \pi_F, \mathcal{M})$ is an associated fibre bundle then its cross-sections are in bijective correspondence with maps $\phi : P(\xi) \to F$ that satisfy*

$$\phi(pg) = g^{-1}\phi(p) \text{ for all } p \in P(\xi) \text{ and } g \in G. \tag{5.3.19}$$

*The cross-section $s_\phi$ corresponding to such a map $\phi$ is defined by*

$$s_\phi(x) := [p, \phi(p)] \tag{5.3.20}$$

*where $p$ is any point in $\pi^{-1}(\{x\})$.*

**Proof**

Given $\phi$, define $s_\phi$ by Eq. (5.3.20). This is independent of the choice of $p \in \pi^{-1}(\{x\})$ since $[pg, \phi(pg)] = [pg, g^{-1}\phi(p)] = [p, \phi(p)]$. Furthermore, $\pi_F(s_\phi(x)) = \pi(p) = x$, and hence $s_\phi$ is a cross-section.

Conversely, let $s : \mathcal{M} \to P_F$ be a cross-section and define $\phi_s : P(\xi) \to F$ by

$$\phi_s(p) := \iota_p^{-1}(s(x)) \tag{5.3.21}$$

where $p \in \pi^{-1}(\{x\})$ and $\iota_p : F \to \pi_F^{-1}(\{x\})$ is the injection of $F$ into $P_F$ (and onto $\pi_F^{-1}(\{x\})$) defined previously by $\iota_p(v) := [p, v]$.

Now, $\iota_p(v) = [p, v] = [pg, g^{-1}v] = \iota_{pg}(g^{-1}v)$, and hence $\phi_s(pg) = \iota_{pg}^{-1}(s(x)) = \iota_{pg}^{-1} \circ \iota_p(\phi_s(p)) = g^{-1}\phi_s(p)$, as required.

Finally, note that the maps $\phi \mapsto s_\phi$ and $s \mapsto \phi_s$ are inverses of each other since

$$s_{\phi_s}(x) = [p, \phi_s(p)] = [p, \iota_p^{-1}(s(x))] = s(x) \tag{5.3.22}$$

$$\phi_{s_\phi}(p) = \iota_p^{-1}(s_\phi(x)) = \iota_p^{-1}([p, \phi(p)]) = \phi(p). \tag{5.3.23}$$

**Comments**

1. If $\sigma : U \subset \mathcal{M} \to P(\xi)$ is a local trivialising cross-section of $\xi$, the *local representative* $s_U : U \to F$ of a section $s$ of $P_F$ is defined by

$$s_U(x) := \phi_s(\sigma(x)). \tag{5.3.24}$$

Now, Eq. (5.3.22) implies that, for all $x \in U$,

$$s(x) = [\sigma(x), \phi_s(\sigma(x))], \tag{5.3.25}$$

and hence, using the local trivialisations $h : U \times G \to \pi^{-1}(U)$, $h(x,g) := \sigma(x)g$, and $h' : U \times F \to \pi_F^{-1}(U)$, $h'(x,v) := [\sigma(x), v]$, we see that $s(x)$ and $s_U(x)$ are related by

$$h'(x, s_U(x)) = s(x), \tag{5.3.26}$$

for all $x \in U$.

2. If $\sigma_1 : U_1 \to P$ and $\sigma_2 : U_2 \to P$ are two local sections of $P$ with $U_1 \cap U_2 \neq \emptyset$, then there exists some local 'gauge function' $\Omega : U_1 \cap U_2 \to G$ such that $\sigma_2(x) = \sigma_1(x)\Omega(x)$, for all $x \in U_1 \cap U_2$. Then

$$s_{U_2}(x) = \phi_s(\sigma_2(x)) = \phi_s(\sigma_1(x)\Omega(x)) = \Omega(x)^{-1}\phi_s(\sigma_1(x)) \tag{5.3.27}$$

and so the local representatives $s_{U_1}$ and $s_{U_2}$ are related by the gauge transformation

$$s_{U_1}(x) = \Omega(x) s_{U_2}(x), \tag{5.3.28}$$

for all $x \in U_1 \cap U_2$.

Note that the associated trivialisations satisfy $h_1(x,g) = \sigma_1(x)g$ and $h_2(x,g) = \sigma_2(x)g$. Hence

$$h_2^{-1} \circ h_1(x,g) = h_2(\sigma_1(x)g) = h_2^{-1}(\sigma_2(x)\Omega(x)^{-1}g) = (x, \Omega(x)^{-1}g) \tag{5.3.29}$$

and so $\Omega(x)$ is just the transition function $g_{U_1U_2}(x)$ discussed earlier.

## 5.4 Vector Bundles

### 5.4.1 The main definitions

Vector bundles are of considerable importance in theoretical physics because the space of cross-sections of such a bundle carries a natural structure of a vector space; as such, it can play a role analogous to that of the more familiar linear space of functions on a manifold. From one perspective, a vector bundle is simply an associated bundle in which the fibre happens to be a vector space. However, because vector bundles arise so often in practice, it is common to present them in a framework that makes no direct reference to an underlying principal bundle. We shall start this section with this approach; the associated-principal definition will be given at the end.

**Definition 5.15**

1. An $n$-dimensional real (resp. complex) *vector bundle* $(E, \pi, \mathcal{M})$ is a fibre bundle in which each fibre possesses the structure of an $n$-dimensional real (resp. complex) vector space. Furthermore, for each $x \in \mathcal{M}$ there must exist some neighbourhood $U \subset \mathcal{M}$ of $x$ and a local trivialisation $h : U \times \mathbb{R}^n \to \pi^{-1}(U)$ such that, for all $y \in U$, the map $h : \{y\} \times \mathbb{R}^n \to \pi^{-1}(\{y\})$ is linear.

2. A *vector bundle map* between a pair of vector bundles $(E, \pi, \mathcal{M})$ and $(E', \pi', \mathcal{M}')$ is a bundle map $(u, f)$ in which the restriction of $u : E \to E'$ to each fibre is a linear map.

3. The space $\Gamma(E)$ of all cross-sections of a vector bundle $(E, \pi, \mathcal{M})$ is equipped with a natural module structure over the ring $C(\mathcal{M})$ of continuous, real-valued functions on $\mathcal{M}$, defined by:

   (a) $(s_1 + s_2)(x) := s_1(x) + s_2(x)$ for all $x \in \mathcal{M}$; $s_1, s_2 \in \Gamma(E)$

   (b) $(fs)(x) := f(x)s(x)$ for all $x \in \mathcal{M}$; $f \in C(\mathcal{M})$, $s \in \Gamma(E)$.

   In particular, therefore, $\Gamma(E)$ has the structure of a real vector space in which the real numbers are represented by constant functions on $\mathcal{M}$.

In the case where all the spaces are manifolds, the space of cross-sections is a module over the ring $C^\infty(\mathcal{M})$ of smooth functions on the base space $\mathcal{M}$.

**Examples**

1. For any $n$-dimensional vector space $V$, the product space $\mathcal{M}\times V$ is an $n$-dimensional vector bundle over $\mathcal{M}$.

2. The tangent bundle $T\mathcal{M}$ of a differentiable manifold is a real vector bundle whose dimension is equal to the dimension of $\mathcal{M}$. The same is true of the cotangent bundle $T^*\mathcal{M}$.

3. Let $\mathcal{M}$ be a smoothly embedded, $m$-dimensional submanifold of $\mathbb{R}^n$ for some $n$. Then the *normal bundle* of $\mathcal{M}$ in $\mathbb{R}^n$ is an $(n-m)$-dimensional vector bundle $\nu(\mathcal{M})$ over $\mathcal{M}$ with total space

$$E(\nu(\mathcal{M})) := \{(x,\vec{v}) \in \mathcal{M}\times\mathbb{R}^n \mid \vec{v}\cdot\vec{w} = 0, \text{ for all } \vec{w} \in T_x\mathcal{M}\} \quad (5.4.1)$$

and projection map $\pi : E(\nu(\mathcal{M})) \to \mathcal{M}$ defined by $\pi(x,\vec{v}) := x$.

An example is the normal bundle associated with the embedding of the $(n-1)$-sphere in $\mathbb{R}^n$.

4. There is a 'canonical' real line bundle $\gamma_n$ (*i.e.*, a one-dimensional vector bundle) over the projective space $\mathbb{R}P^n$ with total space

$$E(\gamma_n) := \{([x],v) \in \mathbb{R}P^n \times \mathbb{R}^{n+1} \mid v = \lambda x \text{ for some } \lambda \in \mathbb{R}\} \quad (5.4.2)$$

in which $[x]$ denotes the line that passes through $x \in \mathbb{R}^{n+1}$. The projection map is $\pi : E(\gamma_n) \to \mathbb{R}P^n$ defined by $\pi([x],v) := [x]$; thus $\pi^{-1}(\{[x]\})$ is the line in $\mathbb{R}^{n+1}$ that passes through $x$ (and $-x$). Local triviality is demonstrated by noting that if $U \subset S^n$ is any open set that is sufficiently small that it contains no pair of antipodal points, and if $U'$ is the corresponding set in $\mathbb{R}P^n$, then a local homeomorphism $h : U' \times \mathbb{R} \to \pi^{-1}(U')$ can be defined by $h([x],\lambda) := ([x],\lambda x)$.

### 5.4.2 Vector bundles as associated bundles

Let us now show that vector bundles can also be discussed within the associated-principal bundle framework introduced in the last section.

**Theorem 5.10** *Let $\xi = (P, \pi, \mathcal{M})$ be a principal $GL(n, \mathbb{R})$-bundle and let $GL(n, \mathbb{R})$ act on $\mathbb{R}^n$ in the usual way. Then the associated bundle $\xi[\mathbb{R}^n]$ can be given the structure of an $n$-dimensional real vector bundle.*

### Proof

The map $\iota_p : \mathbb{R}^n \to \pi_{\mathbb{R}^n}^{-1}(\{x\})$, $\iota_p(v) := [p, v]$ where $p \in \pi^{-1}(\{x\})$, is a homeomorphism from $\mathbb{R}^n$ onto $\pi_{\mathbb{R}^n}^{-1}(\{x\})$. To give $\pi_{\mathbb{R}^n}^{-1}(\{x\})$ a vector space structure, choose any $p \in \pi^{-1}(\{x\})$ and define

(i) $\iota_p(v_1) + \iota_p(v_2) := \iota_p(v_1 + v_2)$ for all $v_1, v_2 \in \mathbb{R}^n$;

(ii) $\lambda\iota_p(v) := \iota_p(\lambda v)$ for all $v \in \mathbb{R}^n$, for all $\lambda \in \mathbb{R}$.

If $p' \in \pi^{-1}(\{x\})$ is any other choice such that $\iota_{p'}(v') = \iota_p(v)$ for some $v' \in \mathbb{R}^n$ then, for some $g \in GL(n, \mathbb{R})$,

$$\begin{aligned} \iota_{p'}(v_1') + \iota_{p'}(v_2') &= \iota_{p'}(v_1' + v_2') = \iota_{pg}(g^{-1}v_1 + g^{-1}v_2) \\ &= \iota_{pg}(g^{-1}(v_1 + v_2)) = \iota_p(v_1 + v_2) = \iota_p(v_1) + \iota_p(v_2), \end{aligned} \tag{5.4.3}$$

and hence the vector space structure is independent of the choice of $p \in \pi^{-1}(\{x\})$.

To see that this bundle is locally trivial we define a map $h' : U \times \mathbb{R}^n \to \pi_{\mathbb{R}^n}^{-1}(U)$, $h'(x, v) := [h(x, e), v]$ where $h : U \times GL(n, \mathbb{R}) \to \pi^{-1}(U)$ is a trivialising map of $P$ over $U \subset \mathcal{M}$. It is clear that the restriction $h' : \{x\} \times \mathbb{R}^n \to \pi_{\mathbb{R}^n}^{-1}(\{x\})$ of the local trivialising map $h'$ is linear for each $x \in U \subset \mathcal{M}$. **QED**

### Comments

1. More generally, if $V$ is any real, finite-dimensional vector space we can consider the group $GL(V, \mathbb{R})$ of automorphisms of $V$, and the associated bundle $\xi[V]$ is a vector bundle.

In particular, if $\xi$ is the bundle of frames on an $m$-dimensional manifold, the bundle associated with a linear representation of the group $GL(m, \mathbb{R})$ on any vector space $V$ gives rise to a vector bundle. Thus all tensor bundles are vector bundles.

2. The converse statement to the one in the theorem is also true: namely, every vector bundle is bundle isomorphic to an associated bundle of this type.

3. If $(u, f) : \xi \to \xi'$ is a principal $GL(n, \mathbb{R})$-map then $(u_F, f) : \xi[\mathbb{R}^n] \to \xi'[\mathbb{R}^n]$ is a vector bundle map.

4. If $\xi$ is a principal $GL(n, \mathbb{R})$-bundle over $\mathcal{M}$ and $f$ is a function from $\mathcal{M}'$ to $\mathcal{M}$, then the pull-back bundle $f^*(\xi[\mathbb{R}^n])$ admits a natural structure of an $n$-dimensional vector bundle over $\mathcal{M}'$; hence vector bundles pull-back to vector bundles.

# Chapter 6

# Connections in a Bundle

## 6.1 Connections in a Principal Bundle

### 6.1.1 The definition of a connection

An idea of great significance[1] in fibre bundle theory is that of a 'connection', together with the associated concepts of 'parallel transport' and 'covariant differentiation'. In the case of Riemannian geometry on a manifold $\mathcal{M}$, the well-known Christoffel symbol is essentially a connection on the bundle of frames $\mathbf{B}(\mathcal{M})$. For a Yang-Mills theory with a principal bundle $G \to P \to \mathcal{M}$, the Yang-Mills field can be identified with a connection on the bundle space $P$.

The basic idea of parallel transport/covariant differentiation is to compare the points in 'neighbouring' fibres in a way that is not dependent on any particular local bundle trivialisation. This suggests looking for vector fields on the bundle space $P$ that 'point' from one fibre to another. The obvious natural vector fields on $P$ are those that arise from the right action of $G$ on $P$. According to the discussion in Section 4.5, to each element $A \in L(G)$ there corresponds a vector field $X^A$ on $P$(*cf.* Eq. (4.5.1)) which represents the Lie algebra homomorphically in the sense that (Eq. (4.5.27))

$$[X^A, X^B] = X^{[AB]} \quad \text{for all } A, B \in L(G). \tag{6.1.1}$$

[1] A classic reference is Kobayshi & Nomizu (1963).

Furthermore, since the action of $G$ on $P$ is free, the map

$$\iota : L(G) \to \mathrm{VFlds}(P),\ A \mapsto X^A \tag{6.1.2}$$

is an *isomorphism* of $L(G)$ into the Lie algebra of all vector fields on $P$.

However, these particular fields are not suitable for our purposes since they do not point from one fibre to another. On the contrary, the vectors $X_p^A$ are tangent to the fibre at $p \in P$, and hence they point along the fibre, rather than away from it. Technically, $X_p^A$ is said to be a *vertical* vector, *i.e.*, it belongs to the *vertical subspace* $V_pP$ of $T_pP$ defined by

$$V_pP := \{\tau \in T_pP \mid \pi_*\tau = 0\} \tag{6.1.3}$$

where $\pi : P \to \mathcal{M}$ is the projection in the bundle. It is easy to see that the map $A \mapsto X_p^A$ is an isomorphism of $L(G)$ onto $V_pP$; in particular, $\dim V_p = \dim L(G) = \dim G$.

What is needed is some way of constructing vectors that point *away* from the fibre, *i.e.*, elements of $T_pP$ that complement the vertical vectors in $V_pP$. This motivates the following definition of a 'connection'.

**Definition 6.1**

A *connection* in a principal bundle $G \to P \to \mathcal{M}$ is a smooth assignment to each point $p \in P$ of a subspace $H_pP$ of $T_pP$ such that

$$\text{(a) } T_pP \simeq V_pP \oplus H_pP \text{ for all } p \in P; \tag{6.1.4}$$

$$\text{(b) } \delta_{g*}(H_pP) = H_{pg}P \text{ for all } g \in G,\ p \in P \tag{6.1.5}$$

where $\delta_g(p) := pg$ denotes the right action of $G$ on $P$ (the meaning of the word 'smooth' in this context is explained below).

**Comments**

1. The equation (6.1.4) implies that any tangent vector $\tau \in T_pP$ can be decomposed uniquely into a sum of *vertical* and *horizontal*

components lying in $V_pP$ and $H_pP$; these components will be denoted by $\text{ver}(\tau)$ and $\text{hor}(\tau)$ respectively.

Similarly, a vector field $X$ on $P$ gives rise to a pair of vector fields $\text{ver}(X)$ and $\text{hor}(X)$ on $P$ with the property that, for all $p \in P$, $\text{ver}(X)_p \in V_pP$ and $\text{hor}(X)_p \in H_pP$.

The word 'smooth' in the phrase 'smooth assignment' in the definition means that if $X$ is a smooth vector field, then so are $\text{ver}(X)$ and $\text{hor}(X)$.

2. The equation (6.1.5) implies that the vertical–horizontal decomposition of the tangent spaces of $P$ is compatible with the right action of $G$ on $P$.

3. For all $p \in P$, the projection map $\pi : P \to \mathcal{M}$ induces a map $\pi_* : T_pP \to T_{\pi(p)}\mathcal{M}$ whose kernel is $V_pP$. Thus we have the short exact sequence[2]

$$0 \longrightarrow V_pP \longrightarrow T_pP \xrightarrow{\pi_*} T_{\pi(p)}\mathcal{M} \longrightarrow 0. \tag{6.1.6}$$

Then Eq. (6.1.4) shows that $\pi_* : H_pP \to T_{\pi(p)}\mathcal{M}$ is an isomorphism.

4. A connection can be associated with a certain $L(G)$-valued one-form $\omega$ on $P$ in the following way. If $\tau \in T_pP$, define

$$\omega_p(\tau) := \iota^{-1}(\text{ver}(\tau)) \tag{6.1.7}$$

where $\iota$ is the isomorphism of $L(G)$ with $V_pP$ induced by Eq. (6.1.2). Thus:

(a) $\omega_p(X^A) = A \ \forall p \in P, A \in L(G)$ (6.1.8)

(b) $\delta_g{}^*\omega = \text{Ad}_{g^{-1}}\omega, \textit{i.e.}, (\delta_g{}^*\omega)_p(\tau) = \text{Ad}_{g^{-1}}(\omega_p(\tau)), \forall \tau \in T_pP$ (6.1.9)

(c) $\tau \in H_pP$ if and only if $\omega_p(\tau) = 0$. (6.1.10)

5. This introduction of a connection one-form is very useful: in fact, an alternative way of *defining* a connection is to say that it

[2] A sequence of maps and vector spaces $A_1 \xrightarrow{a_1} A_2 \xrightarrow{a_2} \cdots \xrightarrow{a_n} A_{n+1}$ is said to be an *exact* sequence if, for all $i = 1, 2, \cdots n-1$ we have $\text{Ker}(a_{i+1}) = \text{Im}(a_i)$. An exact sequence of the form $0 \xrightarrow{a} B \xrightarrow{b} C \xrightarrow{c} D \longrightarrow 0$ (where 0 denotes the trivial vector space with just one element) is said to be a *short exact sequence*.

is a $L(G)$-valued one-form on $P$ that satisfies Eqs. (6.1.8–6.1.9). A *horizontal* vector is then defined to be one that satisfies Eq. (6.1.10); which leads once more to the direct-sum decomposition of $T_pP$ in Eq. (6.1.4).

6. If $\omega_1$ and $\omega_2$ are a pair of connection one-forms on $P$, then so is the *affine sum* $(f \circ \pi)\omega_1 + (1 - f \circ \pi)\omega_2$ for any smooth function $f \in C^\infty(\mathcal{M})$. In particular this is true for constant functions—*i.e.*, the set of all connection one-forms is a *cone* in the real vector space of all $L(G)$-valued one-forms on $P$.

### 6.1.2 Local representatives of a connection

Now we come to the crucial question of what a connection one-form $\omega$ 'looks' like, and how this is related to Yang-Mills theory.

A Yang-Mills field is usually written in the form $A^a_\mu$, where $\mu$ is a spacetime index and the index '$a$' ranges from $1, 2, \ldots, \dim G$, corresponding to the fact that $A$ transforms according to the adjoint representation of $G$ under rigid (*i.e.*, spacetime independent) gauge transformations. Thus, if we write

$$A(x) := \sum_{\mu=1}^{m} \sum_{a=1}^{\dim G} A^a_\mu(x) E_a \, (dx^\mu)_x \tag{6.1.11}$$

where $\{E_1, E_2, \ldots, E_{\dim G}\}$ is a basis set for $L(G)$, we see that the Yang-Mills field can be regarded—at least locally—as a Lie-algebra valued one-form on $\mathcal{M}$. The precise relation between this field and the one-form connection on the bundle space $P$, is given by the following theorem.

**Theorem 6.1** *Let $\sigma : U \subset \mathcal{M} \to P$ be a local section of a principal bundle $G \to P \to \mathcal{M}$ which is equipped with a connection one-form $\omega$. Define the local $\sigma$-representative of $\omega$ to be the $L(G)$-valued one-form $\omega^U$ on the open set $U \subset \mathcal{M}$ given by $\omega^U := \sigma^*\omega$. Let $h : U \times G \to \pi^{-1}(U) \subset P$ be the local trivialisation of $P$ induced by $\sigma$ according to $h(x, g) := \sigma(x)g$.*

*Then, if $(\alpha, \beta) \in T_{(x,g)}(U \times G) \simeq T_x U \oplus T_g G$ (cf. Eq. (2.3.73)), the local representative $h^*\omega$ of $\omega$ on $U \times G$ can be written in terms of the local 'Yang-Mills' field $\omega^U$ as*

$$(h^*\omega)_{(x,g)}(\alpha, \beta) = Ad_{g^{-1}}(\omega^U_x(\alpha)) + \Xi_g(\beta) \tag{6.1.12}$$

*where $\Xi$ is the Cartan-Maurer $L(G)$-valued one-form on $G$ defined in Eqs. (4.3.11–4.3.12).*

## Proof

Factor the map $h : U \times G \to P$ as

$$\begin{array}{ccccc} U \times G & \stackrel{\sigma\times\mathrm{id}}{\longrightarrow} & P \times G & \stackrel{\delta}{\longrightarrow} & P \\ (x, g) & \mapsto & (\sigma(x), g) & \mapsto & \sigma(x)g \end{array} \tag{6.1.13}$$

Then,

$$\begin{aligned} (h^*\omega)_{(x,g)}(\alpha, \beta) &= ((\sigma \times \mathrm{id})^*\delta^*\omega)_{(x,g)}(\alpha, \beta) = (\delta^*\omega)_{(\sigma(x),g)}(\sigma_*\alpha, \beta) \\ &= \omega_{\sigma(x)g}((\delta \circ i_g)_*\sigma_*\alpha + (\delta \circ j_{\sigma(x)})_*\beta) \end{aligned} \tag{6.1.14}$$

where, by Eq. (2.3.69), $i_g : P \to P\times G$, $p \mapsto (p, g)$, and $j : G \to P\times G$, $g \mapsto (p, g)$, so that

$$\delta \circ i_g(p) = \delta(p, g) = pg, \ \textit{i.e.}, \ \delta \circ i_g = \delta_g : P \to P \tag{6.1.15}$$

$$\delta \circ j_p(g) = \delta(p, g) = pg, \ \textit{i.e.}, \ \delta \circ j_p = P_p : G \to P. \tag{6.1.16}$$

Therefore,

$$(h^*\omega)_{(x,g)}(\alpha, \beta) = ({\delta_g}^*\omega_{\sigma(x)g})(\sigma_*\alpha) + \omega_{\sigma(x)g}({P_{\sigma(x)}}_*\beta). \tag{6.1.17}$$

Now, ${\delta_g}^*\omega_{\sigma(x)g} = \mathrm{Ad}_{g^{-1}}(\omega_{\sigma(x)})$, according to the characteristic property Eq. (6.1.9) of a connection one-form. We also know that $\beta = L^A_g$ for some $A \in L(G)$—in fact, $A = \Xi_g(\beta)$ according to the definition of the Cartan-Maurer form given in Eq. (4.3.13). Furthermore, according to Eq. (4.5.11), we have ${P_{\sigma(x)}}_*(L^A_g) = X^A_{\sigma(x)g}$, and of course $\omega(X^A) = A \in L(G)$. Putting together these results with Eq. (6.1.17) we finally obtain:

$$\begin{aligned} (h^*\omega)_{(x,g)}(\alpha, \beta) &= \mathrm{Ad}_{g^{-1}}(\omega_{\sigma(x)}(\sigma_*\alpha)) + \Xi_g(\beta) \\ &= \mathrm{Ad}_{g^{-1}}(\omega^U_x(\alpha)) + \Xi_g(\beta) \end{aligned} \tag{6.1.18}$$

for all $(\alpha, \beta) \in T_x U \oplus T_g G$. **QED**

### 6.1.3 Local gauge transformations

The discussion above shows us that a connection one-form $\omega$ can be decomposed locally as the sum of a Yang-Mills field on $\mathcal{M}$ plus a fixed $L(G)$-valued one-form on $G$. Hence—at least locally—specifying a connection is equivalent to giving a Yang-Mills field.

This raises the interesting question of how the familiar gauge transformations arise in the fibre bundle formalism. There are two answers, whose subtle difference depends on whether the gauge group is thought of in an active way or a passive way. In general, a *gauge transformation* in the principal bundle $G \to P \to \mathcal{M}$ is defined to be any principal automorphism of the bundle. If $\phi : P \to P$ is such a map, and if $\omega$ is a connection on $P$, it is easy to check that $\phi^*(\omega)$ is an $L(G)$-valued one-form that satisfies all the axioms for a connection. The pull-back $\phi^*(\omega)$ is said to be the *gauge transform* of $\omega$ by the gauge transformation $\phi \in \mathrm{Aut}(P)$. This active view is analogous to the way in which a vector field $X$ on a manifold $\mathcal{M}$ can be defined to transform under the action of a diffeomorphism $h : \mathcal{M} \to \mathcal{M}$ as the push-forward $h_*(X)$, with a similar definition for any other tensor field on $\mathcal{M}$.

In the Yang-Mills case, it seems relevant to ask how a local representative $\omega^U$ is affected by such a gauge transformation. We shall answer this question in the course of considering the alternative, 'passive' view of the gauge group. The analogue of this in ordinary differentiable geometry is the relation defined on $x \in U \cap U'$ (see Eq. (3.1.14))

$$X^{\nu'}(x) = \sum_{\mu=1}^{m} X^{\mu}(x)\frac{\partial x^{\nu'}}{\partial x^{\mu}}(x) \tag{6.1.19}$$

between the components $X^{\mu}$ and $X^{\nu'}$ of a vector field with respect to local coordinate charts whose domains are $U$ and $U'$ respectively. This can be related to the 'active' view by noting that if $(U, \phi)$ is a coordinate chart on $\mathcal{M}$, and if $f : \mathcal{M} \to \mathcal{M}$ is a diffeomorphism then a new coordinate chart can be defined as the pair $(f(U), \phi \circ f^{-1})$. If $f$ is close enough to the identity map that $U \cap f(U) \neq \emptyset$, then Eq. (6.1.19) can be applied for all $x$ in this intersection. I shall leave as an exercise [!] the task of investigating the exact relation between the active transformation $X \mapsto f_*(X)$ and the corresponding local coordinate transformation Eq. (6.1.19).

The precise statement for the Yang-Mills case is contained in the following theorem.

**Theorem 6.2** *Let $\omega$ be a connection on the principal bundle $G \to P \to \mathcal{M}$ and let $\sigma_1 : U_1 \to P$ and $\sigma_2 : U_2 \to P$ be two local trivialisations on open sets $U_1, U_2 \subset \mathcal{M}$ such that $U_1 \cap U_2 \neq \emptyset$. Let $A_\mu^{(1)}$ and $A_\mu^{(2)}$ denote the local representatives of $\omega$ with respect to $\sigma_1$ and $\sigma_2$ respectively. Then if $\Omega : U_1 \cap U_2 \to G$ is the unique local gauge function defined by*

$$\sigma_2(x) = \sigma_1(x)\Omega(x), \tag{6.1.20}$$

*the local representatives are related on $U_1 \cap U_2$ by*

$$A_\mu^{(2)}(x) = Ad_{\Omega(x)^{-1}}(A_\mu^{(1)}(x)) + (\Omega^*\Xi)_\mu(x). \tag{6.1.21}$$

**Proof**

If $\partial_\mu$ denotes the local vector field $\frac{\partial}{\partial x^\mu}$ then $A_\mu^{(2)}(x) := (\sigma_2^*\omega)_x(\partial_\mu)$, and using the relation Eq. (6.1.20) to factorise the local cross-section $\sigma_2 : U_1 \cap U_2 \to P$ as

$$\begin{array}{ccccc} U_1 \cap U_2 & \xrightarrow{\sigma_1 \times \Omega} & P \times G & \xrightarrow{\delta} & P \\ x & \mapsto & (\sigma_1(x), \Omega(x)) & \mapsto & \sigma_1(x)\Omega(x) \end{array} \tag{6.1.22}$$

we get

$$A_\mu^{(2)}(x) = ((\sigma_1 \times \Omega)^*\delta^*\omega)_x(\partial_\mu) = (\delta^*\omega)_{(\sigma_1(x),\Omega(x))}(\sigma_{1*}(\partial_\mu)_x, \Omega_*(\partial_\mu)_x) \tag{6.1.23}$$

which, by the same type of argument used in the proof of Eq. (6.1.12), can be rewritten as

$$\begin{aligned} A_\mu^{(2)}(x) &= \mathrm{Ad}_{\Omega(x)^{-1}}(A_\mu^{(1)}(x)) + \Xi_{\Omega(x)}(\Omega_*(\partial_\mu)_x) \\ &= \mathrm{Ad}_{\Omega(x)^{-1}}(A_\mu^{(1)}(x)) + (\Omega^*\Xi)_x(\partial_\mu). \end{aligned} \tag{6.1.24}$$

**QED**

**Corollary**

If $G$ is a matrix group, this result can be rewritten using Eq. (4.3.16) as

$$A_\mu^{(2)}(x) = \Omega(x)^{-1} A_\mu^{(1)}(x)\Omega(x) + \Omega(x)^{-1}\partial_\mu\Omega(x). \qquad (6.1.25)$$

**Comments**

1. Eq. (6.1.25) looks just like the familiar Yang-Mills gauge transformation which, in a sense, it is. But note that it relates a *pair* of local Yang-Mills fields whose regions of definition in $\mathcal{M}$ are typically different (albeit overlapping)—it does not refer to a single Yang-Mills field.

2. The relation with the transformation for a single field is as follows. If $\sigma : U \to P$ is a local section of $G \to P \to \mathcal{M}$ with $A := \sigma^*(\omega)$, then an active gauge transformation $\phi : P \to P$ induces a transformation $A \mapsto \sigma^*(\phi^*\omega) = (\phi \circ \sigma)^*\omega$. But there exists some $\Omega : U \to G$ such that, for all $x \in U$, $\sigma(x) = \phi \circ \sigma(x)\Omega(x)$, and then the argument used in the proof of the theorem shows that the transformation of the local representative $A$ can be written in the familiar form:

$$A_\mu(x) \mapsto \Omega(x) A_\mu(x)\Omega^{-1}(x) + \Omega(x)\partial_\mu\Omega(x)^{-1}. \qquad (6.1.26)$$

Note that, if the bundle is trivial, a cross-section $\sigma$ can be defined on all of $\mathcal{M}$, and then Eq. (6.1.26) refers to a globally-defined $L(G)$-valued one-form on $\mathcal{M}$.

3. If the principal bundle is non-trivial, it is not possible to describe the connection $\omega$ in terms of a *single* Yang-Mills field on $\mathcal{M}$. Instead, one must cover $\mathcal{M}$ with local trivialising charts, and then the local Yang-Mills fields associated with any pair of overlapping charts $U_i, U_j$ will be related on $U_i \cap U_j$ by Eq. (6.1.25) with the corresponding local gauge function $\Omega_{ij}(x)$ satisfying the relation $\sigma_i(x) = \sigma_j(x)\Omega_{ij}(x)$. Note that these functions $\Omega_{ij} : U_i \cap U_j \to G$ are precisely the bundle transition functions discussed in Section 5.3.2.

4. Conversely, if $\{U_i\}_{i\in I}$ is some indexed set of trivialising open sets that cover $\mathcal{M}$ with corresponding local sections $\sigma_i : U_i \to P$ and

transition functions $\Omega_{ij} : U_i \cap U_j \to G$, a 'patching' argument shows that a unique connection one-form on $P$ is defined by any set of local $L(G)$-valued one-forms $A^{(i)}$ on $U_i \subset \mathcal{M}$ that are related pairwise by Eq. (6.1.25).

### 6.1.4 Connections in the frame bundle

An interesting example of the constructions above is a connection in the principal $GL(m, \mathbb{R})$-bundle $\mathbf{B}(\mathcal{M})$ of frames on an $m$-dimensional manifold $\mathcal{M}$. Any local coordinate chart $(U, \phi)$ on $\mathcal{M}$ provides a local section $\sigma : U \to \mathbf{B}(\mathcal{M})$ by associating with $x \in U \subset \mathcal{M}$ the local frame $((\partial/\partial x^1)_x, \ldots, (\partial/\partial x^m)_x)$. If $\omega$ is a connection one-form on $\mathbf{B}(\mathcal{M})$ let $\Gamma := \sigma^*\omega$ denote the associated $L(GL(m, \mathbb{R}))$-valued one-form on $U$, and consider the relation between $\Gamma$ and the local one-form $\Gamma'$ associated with another coordinate chart $(U', \phi')$ such that $U \cap U' \neq \emptyset$.

The local frame associated with the local section $\sigma'$ of $\mathbf{B}(\mathcal{M})$ is $x \mapsto ((\partial/\partial x'^1)_x, \ldots, (\partial/\partial x'^m)_x)$, and the transition function $J : U \cap U' \to GL(m, \mathbb{R})$ is just the Jacobian of the coordinate transformations:

$$(\partial_{\mu'})_x = \sum_{\nu=1}^{m} J^\nu_\mu(x)(\partial_\nu)_x, \quad \text{for all } x \in U \cap U' \qquad (6.1.27)$$

$$J^\nu_\mu(x) := \frac{\partial x^\nu}{\partial x'^\mu}(x). \qquad (6.1.28)$$

Then

$$\begin{aligned} \Gamma'_\mu(x) &:= (\sigma'^*\omega)_x(\partial_{\mu'}) = \sum_{\alpha=1}^{m} J^\alpha_\mu(x)(\sigma'^*\omega)_x(\partial_\alpha) \\ &= \sum_{\alpha=1}^{m} J^\alpha_\mu(x)\big(J^{-1}(x)\Gamma_\alpha(x)J(x) + J^{-1}(x)\partial_\alpha J(x)\big) \qquad (6.1.29) \end{aligned}$$

where we have used the general result Eq. (6.1.25) for a connection in a bundle whose structure group is a matrix group.

If $\{G^\lambda_\chi \mid \chi, \lambda = 1, 2, \ldots, m\}$ is some basis for the Lie algebra $M(m, \mathbb{R})$ (the set of all $m \times m$ real matrices) of $GL(m, \mathbb{R})$, we can write the matrix-valued one-form $\Gamma_\mu$ as $(\Gamma_\mu)^\epsilon_\delta = \sum_{\lambda,\chi=1}^m \Gamma_{\mu\lambda}{}^\chi (G^\lambda_\chi)^\epsilon_\delta$. In particular, if we pick the natural basis set $(G^\lambda_\chi)^\epsilon_\delta := \delta^\epsilon_\chi \delta^\lambda_\delta$ then Eq. (6.1.29) becomes the well-known transformation law for the components $\Gamma_{\mu\delta}{}^\epsilon$ of an affine connection on $\mathcal{M}$:

$$\Gamma'_{\mu\delta}{}^\epsilon(x) = \sum_{\alpha,\rho=1}^m \frac{\partial x^\alpha}{\partial x'^\mu}\frac{\partial x^\rho}{\partial x'^\delta}\frac{\partial x'^\epsilon}{\partial x^\chi}\Gamma_{\alpha\rho}{}^\chi(x) + \sum_{\lambda=1}^m \frac{\partial x'^\epsilon}{\partial x^\lambda}\frac{\partial^2 x^\lambda}{\partial x'^\mu \partial x'^\delta}. \quad (6.1.30)$$

Note that, as in the transformation Eq. (6.1.25) in the analogous case of Yang-Mills theory, the transformation in Eq. (6.1.30) is only defined for points $x \in \mathcal{M}$ that lie in the intersection of the two coordinate charts involved. In particular, Eq. (6.1.30) is valid globally only if there is a globally-defined coordinate chart, which for a generic manifold is not the case.

## 6.2 Parallel Transport

### 6.2.1 Parallel transport in a principal bundle

We can now describe what is meant by 'parallel transport' in a principal bundle $\xi = (P, \pi, \mathcal{M})$ equipped with a connection $\omega$. First, there is a technique for generating horizontal vector fields, *i.e.*, fields whose flow lines move from one fibre into another.

**Definition 6.2**

Since $\pi_* : H_pP \to T_{\pi(p)}\mathcal{M}$ is an isomorphism, to each vector field $X$ on $\mathcal{M}$ there exists a unique vector field, denoted $X^\uparrow$, on $P$ such that, for all $p \in P$,

(a) $\pi_*(X^\uparrow_p) = X_{\pi(p)}$

(b) $\mathrm{ver}(X^\uparrow_p) = 0$.

This vector field $X^\uparrow$ is known as the *horizontal lift* of $X$.

**Comments**

1. The act of horizontal lifting is $G$-equivariant in the sense that $\delta_{g*}(X_p^\uparrow) = X_{pg}^\uparrow$.

2. A sufficient condition for a vector field $Y$ on $P$ to be the horizontal lift of a field on $\mathcal{M}$ is that

(a) $\text{ver}(Y) = 0$; and

(b) $\delta_{g*}(Y_p) = Y_{pg}$ for all $p \in P$, and $g \in G$

since we can then define unambiguously $X_x := \pi_*(Y_p)$ for any $p \in \pi^{-1}(\{x\})$. Then $Y = X^\uparrow$.

3. Horizontal lifting has several pleasant algebraic properties:

(a) $(X^\uparrow + Y^\uparrow) = (X + Y)^\uparrow$

(b) If $f \in C^\infty(\mathcal{M})$, then $(fX)^\uparrow = f \circ \pi\, X^\uparrow$

(c) Using the general rule Eq. (3.1.31) for '$h$-relatedness' we see that

$$\pi_*(\text{hor}[X^\uparrow, Y^\uparrow]) = \pi_*([X^\uparrow, Y^\uparrow]) = [\pi_* X^\uparrow, \pi_* Y^\uparrow] = [X, Y] \tag{6.2.1}$$

from which it follows that

$$[X, Y]^\uparrow = \text{hor}([X^\uparrow, Y^\uparrow]). \tag{6.2.2}$$

**Definition 6.3**

Let $\alpha$ be a smooth curve that maps a closed interval $[a, b] \subset \mathbb{R}$ into $\mathcal{M}$ (*i.e.*, $\alpha$ is the restriction to $[a, b]$ of a smooth curve defined on some open interval containing $[a, b]$). A *horizontal lift* of $\alpha$ is a curve $\alpha^\uparrow : [a, b] \to P$ which is horizontal (*i.e.*, $\text{ver}[\alpha^\uparrow] = 0$) and such that $\pi(\alpha^\uparrow(t)) = \alpha(t)$ for all $t \in [a, b]$.

The key to understanding parallel translation is the following theorem concerning the existence of horizontal lifts of curves:

**Theorem 6.3** *For each point $p \in \pi^{-1}\{\alpha(a)\}$, there exists a unique horizontal lift of $\alpha$ such that $\alpha^\uparrow(a) = p$.*

### Proof

Extend $\alpha$ to the open interval $I := (a-\epsilon, b+\epsilon)$ for some $\epsilon > 0$. Then $\alpha^*(\xi)$ is a principal $G$-bundle over $I$, and ${\alpha_\xi}^*(\omega)$ is a connection one-form on $P(\alpha^*(\xi))$ [Exercise!].

$$\begin{array}{ccc} P(\alpha^*(\xi)) & \xrightarrow{\alpha_\xi} & P \\ \downarrow & & \downarrow \pi \\ I & \xrightarrow{\alpha} & \mathcal{M} \end{array} \qquad \alpha_\xi(t,p) := p. \tag{6.2.3}$$

Let $\beta : I \to P(\alpha^*(\xi))$ be the integral curve that passes through $(a,p) \in P(\alpha^*(\xi)) \subset I \times P$ of the unique horizontal lift of the vector field $d/dt$ on $I$. Define $\alpha^\uparrow := \alpha_\xi \circ \beta$. Clearly $\pi(\alpha^\uparrow(t)) = \alpha(t)$ (so that $\alpha^\uparrow(t) \in P(\alpha^*(\xi))$) and

$$\omega_{\alpha^\uparrow(t)}([\alpha^\uparrow]) = \omega_{\alpha^\uparrow(t)}([\alpha_\xi \circ \beta]) = \omega_{\alpha^\uparrow(t)}(\alpha_{\xi*}[\beta]) = ({\alpha_\xi}^*\omega)_{\beta(t)}([\beta]) = 0 \tag{6.2.4}$$

where the last equality follows since ${\alpha_\xi}^*w$ is a connection one-form on $P(\alpha^*(\xi))$, and $[\beta]$ is horizontal (because $\beta$ is the integral curve of a horizontal vector field). Thus $\alpha^\uparrow$ is horizontal.

If $\gamma : [a,b] \to P$ is any other horizontal lift of $\alpha$, it can be pulled back by $\alpha_\xi$ to give the curve in $P(\alpha^*(\xi))$, $t \mapsto (\alpha_\xi^{-1} \circ \gamma)(t) := (t, \gamma(t))$ with $\alpha_\xi(t,\gamma(t)) = \gamma(t)$. Hence $({\alpha_\xi}^*\omega)([\alpha_\xi^{-1} \circ \gamma]) = 0$, and so the lift $t \mapsto (t,\gamma(t))$ to $P(\alpha^*(\xi))$ of the curve $t \mapsto t$ in $I$ is horizontal. But this is clearly unique and equal to $\beta$. Therefore $\alpha^\uparrow$ is unique. **QED**

### Comments

1. In order to make full use of the ideas above (for example, when defining covariant derivatives) it is necessary to have as explicit an expression as possible for the horizontal lift $\alpha^\uparrow$ of a curve $\alpha$ in terms of the connection one-form $\omega$. In practice, this usually means making a comparison between the horizontal lift $\alpha^\uparrow$ and some other 'natural', but non-horizontal, lift of $\alpha$.

Suppose that $\beta : [a,b] \to P$ is such a curve, *i.e.*, $\pi(\beta(t)) = \alpha(t)$ for all $t \in [a,b]$. Then there exists some unique function $g : [a,b] \to G$

such that, for all $t \in [a,b]$,

$$\alpha^{\uparrow}(t) = \beta(t)g(t). \tag{6.2.5}$$

This relation can be factored as

$$\begin{array}{ccccc} [a,b] & \xrightarrow{\beta\times g} & P\times G & \xrightarrow{\delta} & P \\ t & \mapsto & (\beta(t),g(t)) & \mapsto & \beta(t)g(t) \end{array} \tag{6.2.6}$$

and then

$$\begin{aligned} [\alpha^{\uparrow}]_{\beta(t)g(t)} &= [\delta\circ(\beta\times g)]_{\beta(t)g(t)} = \delta_*([\beta],[g])_{(\beta(t),g(t))} \\ &= (\delta\circ i_{g(t)})_*[\beta]_{\beta(t)} + (\delta\circ j_{\beta(t)})_*[g]_{g(t)}. \end{aligned} \tag{6.2.7}$$

Then using the type of argument following Eq. (6.1.17) and the fact that $\omega([\alpha^{\uparrow}]) = 0$, we find

$$0 = \mathrm{Ad}_{g(t)^{-1}}(\omega_{\beta(t)}([\beta])) + \Xi_{g(t)}([g]). \tag{6.2.8}$$

But $[g]_{g(t)} = g_*(d/dt)_t$, and $\Xi_{g(t)}(g_*(d/dt)_t) = (g^*\Xi)_t(d/dt)_t$. In particular, if $G$ is a group of matrices, Eq. (6.2.8) can be rewritten as

$$0 = g(t)^{-1}\omega_{\beta(t)}([\beta])g(t) + g(t)^{-1}\frac{dg}{dt}(t). \tag{6.2.9}$$

This is the differential equation that determines the horizontal-lift function $g(t)$ in terms of the connection $\omega$.

2. If $\sigma : U \to P$ is a local section of $\xi$ associated with a coordinate chart $U \subset \mathcal{M}$, a natural choice for a lift of a curve $\alpha : [a,b] \to U$ is $\beta(t) := \sigma(\alpha(t))$. Then $[\beta] = \sigma_*[\alpha]$, and hence $\omega_{\beta(t)}([\beta]) = (\sigma^* w)_{\alpha(t)}([\alpha])$. But $\sigma^*\omega$ is the local representative $\omega^U$, which we have identified earlier with a Yang-Mills field $A$. In terms of this field, Eq. (6.2.9) becomes

$$0 = \sum_{\mu=1}^{m} g(t)^{-1}A_\mu(\alpha(t))g(t)\frac{dx^\mu(\alpha(t))}{dt} + g(t)^{-1}\frac{dg(t)}{dt} \tag{6.2.10}$$

where $x^\mu$ are the local coordinates in this chart. If the boundary conditions on $t \mapsto g(t)$ are $g(a) = g_0 \in G$, Eq. (6.2.10) can be re-expressed as the matrix-valued integral equation

$$g(t) = g_0 - \int_a^t ds\, A_\mu(\alpha(s))\, \dot{\alpha}^\mu(s)g(s) \tag{6.2.11}$$

where we have written the components of the tangent vector $[\alpha]$ as $\dot\alpha^\mu(t) := dx^\mu(\alpha(t))/dt$, and where summation of the dummy index $\mu$ is understood.

This integral equation for $g(t)$ can be solved as the path-ordered integral

$$g(t) = \Big(\mathbf{P}\exp - \int_a^t ds\, A_\mu(\alpha(s))\,\dot\alpha^\mu(s)\Big)g_0 := \Big(1 - \int_a^t ds\, A_\mu(\alpha(s))\dot\alpha^\mu(s) + \int_a^t ds_1 \int_a^{s_1} ds_2\, A_{\mu_1}(\alpha(s_1))A_{\mu_2}(\alpha(s_2))\dot\alpha^{\mu_1}(s_1)\dot\alpha^{\mu_2}(s_2) + \cdots\Big)g_0. \tag{6.2.12}$$

Thus the final local expression for the horizontal lift $\alpha^\uparrow$ of the curve $\alpha : [a,b] \to U \subset \mathcal{M}$ is

$$\alpha^\uparrow(t) = \sigma(\alpha(t))\Big(\mathbf{P}\exp - \int_a^t ds\, A_\mu(\alpha(s))\,\dot\alpha^\mu(s)\Big)g_0. \tag{6.2.13}$$

3. If $A$ is subjected to a gauge transformation of the type in Eq. (6.1.26) then it is fairly easy to see [Exercise!] that the path-ordered integral transforms homogeneously as

$$\mathbf{P}\exp - \int_a^t ds\, A_\mu(\alpha(s))\,\dot\alpha^\mu(s) \mapsto \Omega(\alpha(t))^{-1}\Big(\mathbf{P}\exp - \int_a^t ds\, A_\mu(\alpha(s))\,\dot\alpha^\mu(s)\Big)\Omega(\alpha(a)). \tag{6.2.14}$$

4. If $G$ is a group of matrices, and if $\alpha$ is a *closed loop* (so that $\alpha(a) = \alpha(b)$), then Eq. (6.2.14) shows that

$$W_\alpha[A] := \operatorname{tr}\Big(\mathbf{P}\exp - \oint_\alpha ds\, A_\mu(\alpha(s))\,\dot\alpha^\mu(s)\Big) \tag{6.2.15}$$

is a *gauge-invariant* function of the gauge field $A$. It is known as the *Wilson loop* function and plays an important role in many approaches to the quantisation of gauge fields. The analogous object in general relativity is a central ingredient in the 'loop-variable' approach to the canonical quantisation of gravity. □

We can now give a precise meaning to the concept of parallel translation:

**Definition 6.4**

> Let $\alpha : [a,b] \to \mathcal{M}$ be a curve in $\mathcal{M}$. The *parallel translation* along $\alpha$ is the map $\tau : \pi^{-1}(\{\alpha(a)\}) \to \pi^{-1}(\{\alpha(b)\})$ obtained by associating with each point $p \in \pi^{-1}(\{\alpha(a)\})$ the point $\alpha^{\uparrow}(b) \in \pi^{-1}(\{\alpha(b)\})$ where $\alpha^{\uparrow}$ is the unique horizontal lift of $\alpha$ that passes though $p$ at $t = a$.

**Comments**

1. Since a horizontal curve is mapped into a horizontal curve by the right action $\delta$ of $G$ on $P$ we have, for all $g \in G$, $\tau \circ \delta_g = \delta_g \circ \tau$. In particular, if $p_1 = \tau(p)$ then $p_1 g = \tau(p)g = \tau(pg)$, which implies that $\tau$ is a bijection of fibres.

2. As remarked already, an interesting case is when $\alpha$ is a closed curve (*i.e.*, a loop) in $\mathcal{M}$. There is no reason why the horizontal lift should also be closed, and in general we get a non-trivial map from $\pi^{-1}(\{\alpha(a)\})$ onto itself given by

$$p \mapsto p\left(\mathbf{P}\exp - \oint_\alpha ds\, A_\mu(\alpha(s))\dot{\alpha}^\mu(s)\right). \tag{6.2.16}$$

Thus we have a natural map from the loop space of $\mathcal{M}$ into $G$. The subgroup of all elements in $G$ that can be obtained in this way is called the *holonomy group* of the bundle at the point $\alpha(0) \in \mathcal{M}$; it plays an important role in understanding the relation between the connection and certain topological properties of $\mathcal{M}$.

### 6.2.2 Parallel transport in an associated bundle

The next important step is to extend the ideas of connections and parallel transport to associated fibre bundles. This is performed by means of the following definition of the 'vertical' and 'horizontal' subspaces of a tangent space to the associated bundle.

**Definition 6.5**

1. Let $\omega$ be a connection in the principal $G$-bundle $\xi = (P, \pi, \mathcal{M})$, and let $\xi[F] = (P_F, \pi_F, \mathcal{M})$ be the bundle associated to $\xi$ via a left action of $G$ on $F$. The *vertical subspace* of the tangent space $T_y(P_F)$, $y \in P_F$, is defined as (*cf.* Eq. (6.1.3))

$$V_y(P_F) := \{\tau \in T_y(P_F) \mid \pi_{F*}\tau = 0\}. \tag{6.2.17}$$

2. Let $k_v : P(\xi) \to P_F$, $v \in F$, be defined by $k_v(p) := [p, v]$. Then the *horizontal subspace* of the tangent space $T_{[p,v]}(P_F)$ is defined as

$$H_{[p,v]}(P_F) := k_{v*}(H_pP). \tag{6.2.18}$$

**Comments**

1. Since $k_{g^{-1}v} \circ \delta_g = k_v$, the definition of $H_{[p,v]}(P_F)$ is independent of the choice of elements $(p, v)$ in the equivalence class of $y = [p, v] \in P_F$.

2. Let $\alpha : [a, b] \to \mathcal{M}$ and let $[p, v]$ be any point in $\pi_F^{-1}(\{\alpha(a)\})$. Let $\alpha^\uparrow$ be the unique horizontal lift of $\alpha$ to $P(\xi)$ such that $\alpha^\uparrow(a) = p$. Then the curve

$$\alpha_F^\uparrow(t) := k_v(\alpha^\uparrow(t)) = [\alpha^\uparrow(t), v] \tag{6.2.19}$$

is the horizontal lift of $\alpha$ to $P_F$ that passes through $[p, v]$ at $t = a$. This leads to the concept of *parallel translation* (or *transportation*) in the associated bundle as the map $\tau_F : \pi_F^{-1}(\{\alpha(a)\}) \to \pi_F^{-1}(\{\alpha(b)\})$ obtained by taking each point $y \in \pi_F^{-1}(\{\alpha(a)\})$ into the point $\alpha_F^\uparrow(b)$, where $t \mapsto \alpha_F^\uparrow(t)$ is the horizontal lift of $\alpha$ to $P_F$ that passes through $y$.

3. If $\sigma : U \to P(\xi)$ is a local section of the principal bundle, the natural choice for a lift of $\alpha$ to $P(\xi)$ is $\beta(t) := \sigma(\alpha(t))$, and then $\alpha^\uparrow(t) = \beta(t)g(t)$ where $g : [a, b] \to G$ obeys the differential equation (6.2.10). As usual, this local section generates the trivialisation map $h : U \times G \to \pi^{-1}(U)$, $h(x, g) := \sigma(x)g$, with inverse $u : \pi^{-1}(U) \to U \times G$ defined as $u(p) := (\pi(p), \chi_\sigma(p))$ where $\chi_\sigma : P \to G$ is the

unique map satisfying $p = \sigma(\pi(p))\chi_\sigma(p)$ for all $p \in \pi^{-1}(U)$. Then the image of $\alpha^\uparrow$ in $U \times G$ is

$$u \circ \alpha^\uparrow(t) = (\alpha(t), \chi_\sigma(\beta(t)g(t))) = (\alpha(t), \chi_\sigma(\beta(t)))g(t) = (\alpha(t), g(t)). \tag{6.2.20}$$

Similarly, using the local trivialisation $u' : \pi_F^{-1}(U) \to U \times F$, the image of $\alpha_F^\uparrow(t)$ is the curve

$$t \mapsto (\alpha(t), g(t)v) \in U \times F. \tag{6.2.21}$$

### 6.2.3 Covariant differentiation

The basic problem of defining a derivative of a cross-section $\psi : \mathcal{M} \to P_V$ of a vector bundle is the lack of any natural way of comparing the values of $\psi$ at any pair of neighbouring points in $\mathcal{M}$. More precisely, these values lie in two *different* fibres and although they could be compared by using a local bundle trivialisation, the result would depend on the trivialisation chosen. However, when the bundle is equipped with a connection this can be used to 'pull-back' the fibre over the second point to the fibre over the first, and then a subtraction can be performed unambiguously. The precise definition is as follows.

**Definition 6.6**

Let $\xi = (P, \pi, \mathcal{M})$ be a principal $G$-bundle and let $V$ be a vector space that carries a linear representation of $G$. Let $\alpha : [0, \epsilon] \to \mathcal{M}$, $\epsilon > 0$, be a curve in $\mathcal{M}$ such that $\alpha(0) = x_0 \in \mathcal{M}$, and let $\psi : \mathcal{M} \to P_V$ be a cross-section of the associated vector bundle. Then the *covariant derivative* of $\psi$ in the direction $\alpha$ at $x_0$, is

$$\nabla_\alpha \psi := \lim_{t \to 0} \left( \frac{\tau_V^t \psi(\alpha(t)) - \psi(x_0)}{t} \right) \in \pi_V^{-1}(\{x_0\}) \tag{6.2.22}$$

where $\tau_V^t$ is the (linear) parallel-transport map from the vector space $\pi_V^{-1}(\{\alpha(t)\})$ to the vector space $\pi_V^{-1}(\{x_0\})$.

**Comments**

1. In a local bundle chart, the horizontal lift of $\alpha(t)$ to $U \times V$ that passes though $(x_0, v)$ at $t = 0$ is given by Eq. (6.2.21) as the curve

$t \mapsto (\alpha(t), g(t)v)$ where $g(t)$ satisfies the differential equation Eqs. (6.2.9–6.2.10) with the boundary condition $g(0) = e$. Then, if the local representative of $\psi$ is $\psi_U : U \to V$ satisfying Eq. (5.3.26)—*i.e.*, $h'(x, \psi_U(x)) = \psi(x)$—the element of $V$ that represents $\tau_V^t\psi(\alpha(t))$ is $g(t)^{-1}\psi_U(\alpha(t))$. Thus the local representative of $\nabla_\alpha\psi$ is the element of $V$ given by:

$$\frac{d}{dt}\left(g(t)^{-1}\psi_U(\alpha(t))\right)\bigg|_{t=0} = \left(\frac{d}{dt}(g^{-1}(t))\psi_U(x_0) + g^{-1}(t)\frac{d\psi_U(\alpha(t))}{dt}\right)\bigg|_{t=0}$$
$$= \sum_{\mu=1}^{m} \left(\partial_\mu\psi_U(x_0) + A_\mu(x_0)\right) \frac{dx^\mu(\alpha(t))}{dt}\bigg|_{t=0} \tag{6.2.23}$$

where—in deriving the last expression—we have used (i) the result Eq. (6.2.10) that relates $g(t)$ to the local Yang-Mills field; and (ii) the identity $d(g(t)g(t)^{-1})/dt = 0$.

2. It follows from Eq. (6.2.23) that if the curves $\alpha$ and $\beta$ are tangent at $x_0 \in \mathcal{M}$, then

$$\nabla_\alpha\psi = \nabla_\beta\psi \tag{6.2.24}$$

for all cross-sections $\psi$ of the vector bundle. Thus the following definition is meaningful:

**Definition 6.7**

1. If $v \in T_x\mathcal{M}$, the *covariant derivative* of the section $\psi$ of $P_V$ *along* $v$ is defined to be $\nabla_v\psi := \nabla_\alpha\psi$, where $\alpha$ is any curve in $\mathcal{M}$ that belongs to the equivalence class of $v$.

2. If $X$ is a vector field on $\mathcal{M}$, the *covariant derivative* along $X$ is the linear operator $\nabla_X : \Gamma(P_V) \to \Gamma(P_V)$ on the set $\Gamma(P_V)$ of cross-sections of the vector bundle $P_V$ defined by

$$(\nabla_X\psi)(x) := \nabla_{X_x}\psi. \tag{6.2.25}$$

**Comments**

1. A particular case of Eq. (6.2.25) is $\nabla_\mu := \nabla_{\partial_\mu}$ where—with the aid of Eq. (6.2.23)—Eq. (6.2.25) becomes the familiar expression

$$(\nabla_\mu\psi)(x) = \partial_\mu\psi(x) + A_\mu(x)\psi(x). \tag{6.2.26}$$

2. Using the notation in Eq. (5.3.19), the covariant derivative of the cross-section $\psi$ can be expressed in terms of the equivalent function $\phi_\psi : P \to V$ that satisfies $\phi_\psi(pg) = g^{-1}\phi_\psi(p)$ as

$$\nabla_{X_x}\psi = \iota_p(X_x^\uparrow \phi_\psi) \tag{6.2.27}$$

where $\iota$ is the usual map $V \to P_V$, $\iota_p(v) := [p, v]$, and $X^\uparrow$ is the horizontal lift of the vector field $X$ to $P$ [Exercise!]. Thus the function from $P$ to $V$ that corresponds to the cross-section $\nabla_X\psi$ is simply $X^\uparrow(\phi_\psi)$.

3. Not only is $\nabla_X$ a linear operator on $\Gamma(P_V)$, it also possesses a derivation property in the form

$$\nabla_X(f\psi) = f\nabla_X\psi + X(f)\psi \tag{6.2.28}$$

for all $f \in C^\infty(\mathcal{M})$. Note that $\nabla_X$ is also linear in the subscript $X$ (*i.e.*, it is linear with respect to the set of vector fields on $\mathcal{M}$ regarded as a module over $C^\infty(\mathcal{M})$) in the sense that

$$\text{(a)} \quad \nabla_{X+Y}\psi = \nabla_X\psi + \nabla_Y\psi \tag{6.2.29}$$

$$\text{(b)} \quad \nabla_{fX}(\psi) = f\nabla_X\psi. \tag{6.2.30}$$

4. The conditions expressed by Eqs. (6.2.28–6.2.30) are characteristic in the sense that one way of *defining* a covariant derivative on a vector bundle $(P_V, \pi_V, \mathcal{M})$ is as any family of linear maps $\nabla_X : \Gamma(P_V) \to \Gamma(P_V)$, $X \in \text{VFlds}(\mathcal{M})$, that satisfies these conditions. One can then track backwards and eventually arrive at the idea of a connection one-form on the underlying principal bundle $\xi$. □

### 6.2.4 The curvature two-form

All that remains to be done to relate the heuristic theory of Yang-Mills fields to the theory of connections is to explain how the well-known Yang-Mills field strength $F_{\mu\nu}$ arises in this geometrical approach. We know from the heuristic discussion in Yang-Mills theory (which necessarily neglects all global aspects since—in effect—it deals with only a local trivialising chart on $\mathcal{M}$) that $F_{\mu\nu}$ will turn out to be some sort of 'covariant curl' of $A_\mu$ arranged to transform covariantly under the gauge transformation in Eq. (6.1.26), but the discussion above of covariant differentiation is not applicable here since the connection/Yang-Mills field is not a cross-section of a vector bundle but rather a one-form defined on the principal bundle space. The new idea that must be invoked is a special extension of the exterior derivative to incorporate the vertical-horizontal splitting of the tangent spaces.

**Definition 6.8**

1. If $\omega$ is any $k$-form on a principal bundle space $P(\xi)$, the *exterior covariant derivative* of $\omega$ is the horizontal $(k+1)$-form $D\omega$ defined by

$$D\omega := d\omega \circ \text{hor} \tag{6.2.31}$$

   *i.e.*,

$$D\omega(X_1, X_2, \ldots, X_{k+1}) = d\omega(\text{hor}X_1, \text{hor}X_2, \ldots, \text{hor}X_{k+1}) \tag{6.2.32}$$

   for any set $\{X_1, X_2, \ldots, X_{k+1}\}$ of vector fields on $P(\xi)$.

2. If $\omega$ is a connection one-form on $P(\xi)$, the *curvature two-form* of $\omega$ is defined as $G := D\omega$.

The relation of $D\omega$ to the familiar Yang-Mills field strength is contained in the famous *Cartan structural equation*:

**Theorem 6.4** *If $G = D\omega$ is the curvature 2-form of the connection $\omega$, then on an arbitrary pair of vector fields $X$ and $Y$ on $P(\xi)$ we have, for all $p \in P(\xi)$,*

$$G_p(X,Y) = d\omega_p(X,Y) + [\omega_p(X)\omega_p(Y)] \tag{6.2.33}$$

*where $[\omega_p(X)\omega_p(Y)]$ denotes the Lie bracket in $L(G)$ between the Lie algebra elements $\omega_p(X)$ and $\omega_p(Y)$.*

**Proof**

Since both sides of Eq. (6.2.33) are linear functions of $X$ and $Y$ it suffices to prove the relation for the three choices: (i) $X$ and $Y$ both horizontal; (ii) $X$ and $Y$ both vertical; (iii) one of the pair $X$, $Y$ is horizontal, the other is vertical.

(i) If $X$ and $Y$ are both horizontal then $\omega(X) = 0 = \omega(Y)$, and $D\omega(X,Y) = d\omega(X,Y)$. Hence Eq. (6.2.33) is satisfied.

(ii) If $X$ and $Y$ are both vertical then there exist $A, B \in L(G)$ such that $X_p = X_p^A$ and $Y_p = X_p^B$ . Now, according to the definition following Theorem 3.2, a tensor expression like $d\omega_p(X,Y)$ is independent of which local vector fields are chosen to extend $X_p$ and $Y_p$ away from $p \in \mathcal{M}$. In particular, we can evaluate this expression using $X^A$ and $X^B$. Then the right hand side of Eq. (6.2.33) becomes

$$d\omega_p(X^A, X^B) + [\omega_p(X^A)\omega_p(X^B)]. \tag{6.2.34}$$

Now, using the expression Eq. (3.4.15) for the exterior derivative of a two-form we have $d\omega(X^A,X^B) = X^A(\omega(X^B)) - X^B(\omega(X^A)) - \omega([X^A, X^B])$. But $\omega(X^B)$ is the constant Lie algebra element $B$, and hence $X^A(\omega(X^B)) = 0$. Similarly, $X^B(\omega(X^A)) = 0$. By Eq. (4.5.27), $[X^A, X^B] = X^{[AB]}$, and hence $\omega([X^A, X^B]) = \omega(X^{[AB]}) = [AB]$. Thus, since $\omega_p(X^A) = A$ and $\omega_p(X^B) = B$, we see that Eq. (6.2.34) vanishes. However, the left hand side of Eq. (6.2.33) vanishes identically since $X$ and $Y$ are vertical. Thus Eq. (6.2.33) is satisfied.

(iii) If $X$ is horizontal and $Y$ is vertical, then $G(X,Y) = 0$ (because $Y$ is vertical), and $[\omega_p(X)\omega_p(Y)] = 0$ since $\omega(X) = 0$. Thus it remains only to show that $d\omega(X,Y) = 0$. Evaluating this tensorial object at $p \in P$ by invoking the same argument as above, we can replace $Y$ with $X^A$ for some $A \in L(G)$. Now, $X(\omega(X^A)) = 0$ since $\omega(X^A)$ is the constant Lie algebra element $A$; and $\omega(X) = 0$ since $X$ is horizontal. Thus $d\omega_p(X,Y) = -\omega_p([X, X^A])$. However, if $X$ generates the flow $t \mapsto \phi_t^X$ of diffeomorphisms of $\mathcal{M}$, we know from Eq. (4.5.20) that $[X,Y] = \lim_{t\to 0}(Y - \phi_t^X{}_*(Y))/t$. In particular,

$$[X, X^A] = \lim_{t\to 0}(\delta_{\exp tA_*}(X) - X)/t. \tag{6.2.35}$$

Thus, if $X$ is horizontal, so is $[X, X^A]$, and hence $\omega([X, X^A]) = 0$.

**QED**

**Comments**

1. If $\{E_1, E_2, \ldots, E_{\dim G}\}$ is a basis for the Lie algebra $L(G)$, we can write $\omega = \omega^a E_a$, and then Eq. (6.2.33) becomes

$$G^a = d\omega^a + \frac{1}{2} \sum_{b,c=1}^{\dim G} C_{bc}{}^a \omega^b \wedge \omega^c \tag{6.2.36}$$

where $C_{bc}{}^a$ are the structure functions of $L(G)$ (and where summing is understood over the repeated indices). This expression for $G^a$ should be contrasted with the Cartan-Maurer equation Eq. (4.3.10) $d\omega^a + \frac{1}{2}\sum_{b,c=1}^{\dim G} C_{bc}{}^a \omega^b \wedge \omega^c = 0$ for the left-invariant one-forms $\omega^a$ on a Lie group $G$.

2. If $\sigma : U \to P$ is a local section of the principle bundle, the local representative $A := \sigma^*\omega$ of $\omega$ is supplemented with the local representative $F := \sigma^*G$ of the curvature 2-form. It then follows from Eq. (6.2.36) that $F^a = dA^a + \frac{1}{2}\sum_{b,c=1}^{\dim G} C_{bc}{}^a A^b \wedge A^c$. Or, inserting coordinate indices, we get the familiar result

$$F^a_{\mu\nu} = \frac{1}{2}(A^a_{\mu,\nu} - A^a_{\nu,\mu} + \sum_{b,c=1}^{\dim G} C_{bc}{}^a A^b_\mu A^c_\nu). \tag{6.2.37}$$

3. It is easy to prove the *Bianchi identity* $DG = 0$.

4. If $\sigma_1 : U_1 \to P$ and $\sigma_2 : U_2 \to P$ are a pair of local sections with $U_1 \cap U_2 \neq \emptyset$, there exists some local gauge function $\Omega : U_1 \cap U_2 \to G$ such that $\sigma_2(x) = \sigma_1(x)\Omega(x)$. Correspondingly, there are two local representatives for the curvature 2-form $G$—namely $F^{(1)} := \sigma_1{}^*G$ and $F^{(2)} := \sigma_2{}^*G$. Using an analysis very similar to that employed in the derivation of the gauge relation Eqs. (6.1.21–6.1.25), it can be shown [Exercise!] that these curvature representatives are related by

$$F^{(2)}_{\mu\nu}(x) = \Omega(x)^{-1} F^{(1)}_{\mu\nu}(x)\Omega(x) \tag{6.2.38}$$

for all $x \in U_1 \cap U_2$. □

This completes the derivation of the basic relation between the mathematical theory of connections in principal and associated bundles, and the physicists' familiar theory of the Yang-Mills field and its gauge transformations. Note that, although we have talked above about the Yang-Mills field, the same analysis applies also to the Riemannian connection in the $GL(m, \mathbb{R})$-bundle of frames $\mathbf{B}(\mathcal{M})$. In this case, the parallel transport and covariant derivatives coincide with the familiar operations from elementary Riemannian geometry, and the curvature 2-form that takes its values in the Lie algebra of $GL(m, \mathbb{R})$ is nothing but the usual curvature tensor in a non-holonomic basis.

# Bibliography


Abraham, R. & Marsden, J. (1980), *Foundations of Mechanics*, Benjamin, London.

Beltrametti, E. & Cassinelli, G. (1981), *The Logic of Quantum Mechanics*, Addison-Wesley, London.

Bott, R. & Tu, L. (1982), *Differential Forms in Algebraic Topology*, Springer-Verlag, New York.

Bourbaki (1966), *Elements of Mathematics: General Topology*, Addison-Wesley, London.

Chevalley, C. (1946), *Theory of Lie Groups*, Princeton University Press, Princeton.

Császár, A. (1978), *General Topology*, Adam Hilger, London.

Dugundji, J. (1996), *Topology*, Allyn and Bacon, Boston.

Fröhlich, O. (1964), *Math. Ann.* **156**, 79.

Hawking, S., King, A. & McCarthy, P. (1976), *J. Math. Phys.* **17**, 171.

Helgason, S. (1962), *Differential Geometry and Symmetric Spaces*, Academic Press, New York.

Hicks, N. (1965), *Notes on Differential Geometry*, Van Nostrand, New York.

Husemoller, D. (1966), *Fibre Bundles*, McGraw-Hill, New York.

Isham, C. (1984), Topological and global aspects of quantum theory, *in* B. DeWitt & R. Stora, eds, 'Relativity, Groups and Topology II', North-Holland, Amsterdam, pp. 1062–1290.

Isham, C. (1995), *Lectures on Quantum Theory: Mathematical and Structural Foundations*, Imperial College Press, London.

Johnstone, P. (1986), *Stone Spaces*, Cambridge University Press, Cambridge.

Kelly, J. (1970), *General Topology*, Van Nostrand, London.

Kobayshi, S. & Nomizu, K. (1963), *Foundations of Differential Geometry: Volume I*, Wiley, New York.

Kronheimer, E. & Penrose, R. (1967), *Proc. Camb. Phil. Soc.* **63**, 481.

Lang, S. (1972), *Differential manifolds*, Addison Wesley, London.

Larson, R. & Andima, S. (1975), *Jour. Math.* **5**, 177.

Lipschutz, S. (1965), *General Topology*, Schaum Publishing, New York.

MacLane, S. & Moerdijk, I. (1992), *Sheaves in Geometry and Logic: A First Introduction to Topos Theory*, Springer-Verlag, London.

Malament, D. (1977), *J. Math. Phys.* **18**, 1399.

Milnor, J. & Stasheff, J. (1974), *Characteristic Classes*, Princeton University Press.

Montgomery, D. & Zippin, L. (1955), *Topological Transformation Groups*, Interscience, New York.

Postnikov, M. (1986), *Lectures in Geometry: Lie Groups and Lie Algebras*, Mir Publishers, Moscow.

Singer, I. (1978), 'Some remarks on the Gribov ambiguity', *Comm. Math. Phys.* **60**, 7–12.

Sorkin, R. (1991), Spacetime and causal sets, *in* J. D'Olivo, E. Nahmad-Achar, M. Rosenbaum, M. Ryan, L. Urrutia & F. Zertuche, eds, 'Relativity and Gravitation: Classical and Quantum', World Scientific, Singapore, pp. 150–173.

Steenrod, N. (1951), *The Topology of Fibre Bundles*, Princeton University Press, Princeton.

Treves, F. (1967), *Topological Vector Spaces*, Academic Press, London.

Vickers, S. (1989), *Topology via Logic*, Cambridge University Press, Cambridge.

# Index